GW00457263

THE NATURE OF

ARCTIC WHALES

THE NATURE OF

ARCTIC WHALES

Belugas, Bowheads and Narwhals

STEFANI PAINE

GREYSTONE BOOKS
Douglas & McIntyre
Vancouver/Toronto

95 96 97 98 99 5 4 3 2 1

Greystone Books
A division of Douglas & McIntyre Ltd.
1615 Venables Street
Vancouver, British Columbia
V5L 2H1

Published in the United States of America by Sierra Club Books, San Francisco.

Canadian Cataloguing in Publication Data

Paine, Stefani Hewlett, 1946-
The nature of Arctic Whales

"Greystone Books."
Includes bibliographical references and index.
ISBN 1-55054-190-0

1. White whale. 2. Bowhead whale. 3. Narwhal. I. Title
QL737.C4P34 1995 599.5 C95-910087-3

Editing by Nancy Flight
Front jacket photograph by Roy Tanami / Ursus
Back jacket photograph by John K. B. Ford / Ursus
Jacket and text design by Barbara Hodgson
All photographs by John K. B. Ford / Ursus unless otherwise credited
Printed and bound in Singapore by C.S. Graphics Pte. Ltd.

The publisher gratefully acknowledges the assistance of the Canada Council and of the British Columbia Ministry of Tourism, Small Business and Culture for its publishing programs.

PAGE iv: Roy Tanami / Ursus

For Wendy

CONTENTS

FACING PAGE: *This underwater shot of a bowhead whale clearly shows the large mouth opening and white chin markings on this rare baleen whale.* FLIP NICKLIN/MINDEN PICTURES

ACKNOWLEDGEMENTS

A great debt is owed to the whales that I have known for the inspiration they gave me and for the first-hand experience with them that provided much information for this book. I acknowledge the biologists and other researchers who have worked in the North and with the Arctic whales and who published their findings so that I and others might benefit from their knowledge. As always, Dan Goodman and Robert Moshenko of the Canadian Department of Fisheries and Oceans generously provided assistance and information.

I thank my editor, Nancy Flight, for her patience and guidance, as well as David Bruce, Gil Hewlett, Mike Noble of Baron and Associates, Ken Crook, Terry Waines and Wendy Bradley for their help and support, and John Ford for his wonderful photographs and for reviewing the text. I owe great thanks to Roy Tanami, who provided not only time and inspiration but enormous assistance in assembling the unique photographs that illustrate this book. Finally, I acknowledge with sincere gratitude the unfailing encouragement of my husband, Michael Paine.

INTRODUCTION

For years, northern fishermen had probably netted the occasional beluga by accident during their normal fishing operations. But the whales had no commercial value, were a nuisance in the nets and were time-consuming to release, so fishermen avoided them. This all changed on the Pacific coast in the mid-1960s when whales made international news following the accidental capture of killer whales Moby Doll in 1964 and a big male, named Namu, in 1965.

Moby Doll (who was found to be a male on post-mortem) and Namu attracted international media attention. Myths about killer whales as vicious and voracious killers exploded. Oceanariums and aquariums all over North America clamoured for whales of their own. Killer whales, previously viewed with contempt by the fishing community as voracious predators on commercially valuable fish stocks, became a coveted commodity, and some West Coast fishermen-turned-whale-hunters made quick fortunes capturing and selling whales to the new market ready to pay top dollar for a guaranteed drawing card. So in 1967, when a fisherman in Alaska saw belugas in his net, he decided to haul them on board, put them in the hold of his boat and sell them to an aquarium.

The belugas arrived at the Vancouver Aquarium after being held in the boat's hold for two weeks, and what a sight they were, covered in cuts and scrapes. From the day she arrived, the big dirty-white female, the infamous Bela, detested all things human and never surrendered that position. The calf, a male named Lugosi, was covered from head to tail with circular raised welts that bore a suspicious resemblance to ringworm, and his grey-brown skin hung in scabby, ragged chunks. But despite his pathetic physical condition, his attitude was sunny, affectionate and positive. As soon as he connected his human keepers with food, he loved them unconditionally. Ugly, scarred and pocked though he was, he was

FACING PAGE: *A pod of adult male narwhals waiting at the ice edge.*

lively and had a good appetite. In time his skin cleared up, his appetite never lagged, and he eventually grew into a large white beluga.

Twenty-five years ago, little information was available on beluga whale behaviour or feeding in the wild. Thus, care of the two belugas was by trial and error. In those days, not much was known about any whale species, for that matter. Cetology, the study of cetaceans (whales and dolphins) was in its infancy, but the twin events of whales in captivity and the decline of the great whales from excessive hunting came together to attract unprecedented scientific and public interest in whales of all kinds.

The whales ate and they grew enormously, floating like two giant marshmallows, doing nothing to excite, nothing to offend. They were without personality or physical prowess. They never dove. They ate; they slept. It never occurred to anyone that they might be too fat, and therefore too buoyant to dive, but everyone agreed that belugas were boring.

When Bela died in the spring of 1976, the post-mortem revealed a startling truth; inside a 0.3-m (1-foot)-thick covering of blubber, Bela had a trim torso. She had been obese, and obviously Lugosi was too. He was immediately put on a diet, his food cut by half, but nothing happened. His food was cut back even more. He remained as fat as ever, so a different approach was tried. First a young harbour seal and then a dolphin were introduced as playmates. The idea was to get him to exercise, but each time the result was the same. He would chase the new addition for a day or two, then lose interest and lie fat and bored at the surface in his favourite spot, his head shoved into the water inlet.

On a July morning that same year, an unfamiliar sound drew Lugosi away from the water inlet. He felt as much as heard a low, rumbling vibration somewhere nearby. It came from a large, mobile orange crane as it drove slowly around the rear of his pool, stopped, planted its stabilizing pads one at a time like great flat feet outside each wheel and then slowly lowered the tether on the end of its lifting arm into a huge blue box on the back of a large, flat-deck truck. Only the heads and shoulders of five men were visible moving around inside the box. One reached for the hook-ended tether, pulled it into the box and attached it to something inside. Another raised his arm in a wide, circling motion, signalling the machine's operator to raise the crane's lifting arm. A couple of centimetres at a time the arm rose, pulled the tether taut, raising it to reveal four steel cables spread like skinny black talons and holding a 3-m (10-foot)-long white canvas sheet slung between two longer poles. As the sling floated in midair above the box, it turned slightly in the soft morning breeze to reveal the snow-white head of a beluga supported in the sling. A pair of tail flukes fanned delicately from the rear of the sling. Ever so gently the crane arm swung up, over the truck, the driveway and the pool edge to hover over the centre of the pool. Below, Lugosi rolled to one side, one eye riveted to the object overhead. Slowly the canvas sling began its descent to the water's surface.

By now, four neoprene-suited divers bobbed in the centre of the pool, black arms raised, touching the sling poles, holding them apart as the sling and water met and the whale slipped free. From under water emerged a most beautiful female beluga. She was absolutely flawless, milk-white, slim and graceful beyond imagining. The permanent smile of her mouth was drawn as if by an artist with a fine pencil. Slight charcoal grey trimmed her perfect flukes. As she glided effortlessly in the clear blue water, her black eyes surveyed her new surroundings, sweeping over the awestruck mass of floating flesh in the southwest corner. Lugosi didn't move, didn't breathe.

At last he lunged after the magnificent creature, pumping his tail furiously, expanses of sagging fat rippling with the effort. For each strong stroke of her tail, his pumped strenuously five times, and he still couldn't catch her. She saw and ignored his pursuit. Within an hour another whale arrived, this time an adolescent female. The whales were named Kavna and Sanaq, both names derived from Inuit terms for a female spirit of the deep.

In one morning, Lugosi's world turned upside down. His personality blossomed with the arrival of his new companions, and fat poured off him. Released from his blubber prison, he dove, twisted and rolled. Lugosi and the adolescent female, Sanaq, were inseparable, spending hours in rolling, biting, loud chatter, vigorous chases and uninhibited sex play. Their social interactions were intense and complicated; it became obvious that belugas were highly social, with an enormous capacity for play and for physical and vocal communication. They were endlessly alert to each other and everything in their environment. Another myth exploded—belugas were not boring.

In the early spring of the following year, Kavna drew apart from the other two, even to sleep alone, upside down on the bottom of the pool rather than with the other two at the surface. I watched and watched, intensely curious about the change in behaviour. Then one day I knew. I have no idea how or why, but I knew with an absolute certainty that she was pregnant. There was no obvious physical change, and she had never copulated with Lugosi, so she must have mated in the wild the previous spring.

For weeks I watched her through the windows in the cool dark of the underwater viewing room. Her girth expanded enormously, and by the third week of June, I saw lumps move across her huge white belly as the unborn calf turned and stretched. Two long mounds swelled on either side of her vent as her mammaries developed in anticipation of the approaching birth. It was a little like watching a daily soap opera on television; the same characters played out scenes in endless variation as each day moved towards an inexorable but unknown conclusion. The pregnancy would end, but when and how I had no idea. It could end with complications; the calf could be stillborn, or it could be born alive and die within minutes, or the mother could die too, as had happened at an aquarium in the United States some years before. If there were a problem, anaesthesia would be out of the question, since whales are conscious breathers and must be awake to breathe.

At 6:15 P.M. on July 13, Kavna moved quietly to the centre of the pool. By 7:00 P.M. she was clearly in labour. Whatever the outcome, I knew I was about to witness something no one had ever seen before—the underwater birth of a beluga.

At 9:50 P.M., when the skinny, slate-grey calf burst from its mother in a cloud of blood, milk and bubbles, the most powerful impression was the contrast between the fragile, uncoordinated brand-newness of the creature and its tenacious vitality. It exuded an energy and will to survive quite out of proportion to its body.

Watching the calf and his mother over those first few days, I tried to imagine what their lives and the lives of the other two Arctic whales, the narwhal and the bowhead, would be like in the wild. I couldn't imagine how so delicate an animal could ever survive in the wild, yet every Arctic whale begins life thrust from the soothing tropical lagoon of its mother's womb into the shocking cold of Arctic seas.

This book tells the story of the Arctic whales, which follow the ice north as it retreats in the spring and then migrate south again in the fall with the approach of the Arctic winter and its covering of solid ice. Some of what the whales do and where they go is now well understood, but much of their lives remains a tantalizing mystery, shrouded in a world of cold, sea and ice.

PART I

SEASONS OF THE ARCTIC WHALE

Swimming belugas. All three Arctic whale species undertake vast seasonal migrations.

Chapter 1 WINTER

Like a beautiful snow queen, cold rules the Arctic, holding living things at bay with an impenetrable fortress of ice. Yet a few submit to her frosty rule and find a way to live inside the chilly kingdom. In the sea, a trio of whales has conquered the killing cold of the Arctic waters: the snow-white beluga (*Delphinapterus leucas*), dubbed the "sea canary" by early northern whalers; the girthy 80- to 100-tonne (88- to 110-ton) mostly black bowhead whale (*Balaena mysticetus*), the whaler's prize, and the corpse-coloured whale with the bizarre unicorn tusk, the narwhal (*Monodon monoceros*). Other whales might make brief forays into the North during the short Arctic summer, but only these three live their entire lives in and around the ice.

LIVING IN A WORLD OF ICE

If you could stand for a moment on a northern star and look down on the world of the Arctic whales in winter, you would see a huge, white, uneven cape draped over the top of the world. This is the northern ice cap. It floats on the Arctic Ocean and in winter covers an area larger than Canada, more than 12 000 000 km^2 (4.5 million square miles). Half the ice cap is permanent, formed of massive chunks as thick as a two-storey house. The ice moves, cracks and grinds but never disappears. It surrounds gigantic floating ice islands of up to 1000 km^2 (400 square miles) and 165 m (540 feet) thick.

FACING PAGE: *All Arctic whales, including these sounding narwhals, depend on areas of open water in the ice where they can surface to breathe.*

A different kind of ice swirls in slow motion on the perimeter of the permanent polar ice cap. This is the pack ice, or true drift ice. Over the years it constantly forms, grows, melts, freezes and "drifts," as its name implies, eventually to leave the Arctic in an ever-moving ice tongue that stretches 2400 km (1500 miles) out of the Arctic Ocean on the eastern side of Greenland.

The northern ice cap and the pack ice are always there, winter and summer. But winter adds yet another kind of ice, and it is this ice that drives the air-breathing whales to their winter ranges. This is the winter sea ice, also known as the land fast ice because it starts from shore and reaches out to sea from a few metres to several hundred kilometres to join the permanent pack ice.

In the Arctic winter, there is no end to the ice. Ice locks every fiord, surrounds every island and cements itself to every land mass in North America, Eurasia and Greenland that borders the Arctic Ocean. It bonds itself to every shore and invades the land wherever it can find water to freeze. It reaches well below the Arctic Circle (66°33' north latitude) to form a frozen shroud of ice down the east coast of North America along the Labrador coast, around Newfoundland and into the St. Lawrence River. The land fast ice is different from the polar pack or drift ice. The land fast ice is a winter phenomenon and therefore not permanent. Its arrival and departure determines the movements of the Arctic whales.

Like all whales, the Arctic whales are warm-blooded mammals; they must rise to the surface to breathe, and they must give birth to their single young in the water. Unlike all other whales, the Arctic whales can navigate under the ice, find precious breathing holes and maintain body heat year-round in subzero water temperatures. Yet even the Arctic whales must leave most areas in the Arctic for the deepest, darkest, coldest months of the winter. They move to shifting broken ice, or to the subarctic and open water, anywhere where they can surface to breathe, because to stay in the Arctic during winter would be to suffocate under an endless cover of ice.

FACING PAGE: *The water temperature of Arctic seas is below freezing yet remains fluid because of its salt content. It is the water's extraordinary cold that acts as an invisible barrier to all but three special kinds of whale.*

THE CHALLENGE OF COLD WATER

On New Year's Day, some communities undertake a strange ritual known as the Polar Bear Swim. At such events, people of all ages and backgrounds mass at noon on some damp and chilly beach where temperatures hover near freezing. Some of those gathered stand hatted, booted and cloaked against the cold, while others, for no apparent reason, strip to expose a mass of goose flesh. Barefoot and clad only in bathing suits, they stand shivering until some signal is given, whereupon they run cheering into the sea. Cheers become shrieks of anguish as cold water hits sensitive body parts. Soon there are only moans as teeth castanet in uncontrollable chattering. Hands stiff with cold flail at bodies in a fruitless attempt to increase circulation and pound warmth into skin that has lost all sensation. Joints ache with cold. Most "swimmers" leave the water in minutes, and the winner is the one who lasts the longest.

There is a mystery here. The air temperature is ten degrees lower than the water temperature, yet the swimmers suffer far more in the water than in the air. That is because water draws off body heat about twenty times faster than air. Those hardy, or foolhardy, souls who dash in and then out of the water experience, however briefly, what it is to spend time in water that is perhaps 8° or 10°C (46° or 50°F).

If the sea water were cooled another 10°C (18°F) to –1° or –2°C (28° or 31°F), it would not freeze as fresh water would at the same temperature; because of its dissolved salts, it would remain fluid at a lower temperature. This is cold—unimaginably bone-chilling, mind-numbing, killing cold. This is the cold of the Arctic seas.

SIZE AS A DEFENCE AGAINST COLD

The bowhead whale is long and broad, with up to 100 tonnes (110 tons) packed into its 18 m (60 feet) from nose to tail. Its tail flukes span up to 7 m (23 feet) from tip to tip. Beside this black giant, narwhals and belugas are dainty little things, but they are still large animals, averaging 4 to 4.6 m (13 to 15 feet) long and up to 1.6 tonnes (3500 pounds); females are slightly shorter and lighter than males. Arctic whales are big, and so are their newborn offspring. A bowhead calf is already the size of a pickup truck when it enters the world. Narwhal and beluga calves are nearly half as long as their mothers. These whales are big for a reason, and it has to do with keeping warm in cold water.

Whether in air or in water, a small body loses heat faster than a large one because of something known as the ratio of surface to mass. This simply relates how much there is outside (exposed skin) to how much there is inside (fat, muscle, bones and organs).The larger the inside, or the mass, the less surface area there is in relation to the mass and therefore the less surface area from which to lose heat. Thus, it is more energy efficient for a whale to be larger rather than smaller.

It is possible for whales to be so large because their weight is supported by water at all times. In contrast, terrestrial mammals support and move their bulk on legs. That is why no land animal, not even the largest dinosaurs, ever approached the weight of the largest whales.

Whales have marvellously streamlined bodies, further reducing their surface area. There is nothing to impede a smooth passage through the water—no ear lobes, no whiskers, no eyebrows or nose. No lips to speak of, no eyelashes or chin. No wrinkles, folds or jowls. There are no telltale lumps or bumps to indicate the animal's sex, since all sex organs are neatly tucked away except when in use. There are eyes (back and to the sides), a mouth, and that's all. Pectoral (front) fins contain bones reminiscent of the forelimbs and digits of the whales' terrestrial ancestor of more than 60 million years ago, but they look like skin-covered paddles and are used to steer, not swim. To conserve heat, the Arctic whales have reduced their surface area even more than other whales. None has a dorsal (back) fin, and pectoral fins are small.

But reducing the amount of skin exposed to the water is not enough by itself to keep the whales warm and maintain their body temperature at 37°C (98.6°F) in perpetually cold seas. They also need insulation.

Ghostly shapes of belugas at home in their chilly underwater world stay warm wrapped in a thick layer of insulating blubber.

FAT AS INSULATION

You wouldn't consider walking naked to the corner store in winter. Although modesty might play a part, personal comfort is what motivates people to put on appropriate clothing when the temperature dips. We put on a coat to keep warm, but we seldom stop to think of how it actually works—though we know that goose down is warmer than denim and fur is warmer still. This is because the air trapped in the fibres of the fur or goose down (not the fibres themselves) provides an insulating barrier between cold air on the outside and warm air on the inside next to our skin.

If trapped air is such a fine insulator, then perhaps the whales should wear fur coats as other cold-climate animals, such as the arctic fox and polar bear, do. But whales don't, for the same reason that if you are out sailing, warmly dressed for the chilly fall breezes, and you fall overboard, your wet clothing will no longer keep you warm. Once you are wet, cold water draws off body heat whether you have clothes on or not. The whale would have the same problem with wet fur. So, rather than putting insulation on the outside of its skin, the whale puts it on the inside. On us it's called fat; on whales it's called blubber. For a land animal, blubber is a poor solution, because it's heavy. But even though a fat person may have trouble walking to the pool, once in the water, he or she floats! It's the same for the whales. Blubber is a fabulous insulator and a wonderful solution to keeping whales warm in cold water, and its weight disappears in water.

Whereas all whales and dolphins, even tropical species, have some blubber to provide buoyancy and cover muscle, bones and organs to give the animal its smooth torpedo shape, the Arctic whales have taken blubber coverings to the extreme. The bowhead whale wraps its torso in a 30-cm (12-inch)-thick blubber jacket, and the narwhal and beluga have a layer up to 10 cm (4 inches) under their skin. Up to 50 per cent of an Arctic whale's body weight is fat, compared with about 20 per cent of a temperate-water whale's body.

Comfortable and warm in their large, well-insulated bodies, the Arctic whales pass the cold, dark winter far out at sea in open water or in the ever-moving pack ice as they wait for the ice to break in the spring. Exactly where they are and what they do during those dark months remains largely a mystery for the simple reason that we as frail human beings have no such cold-climate adaptations. It is dangerous and expensive for biologists to work in the field in the Arctic winter, especially at sea. Air temperatures average −20°C (−4°F) and drop at times to −50°C (−58°F). Twenty-four-hour darkness persists for months, making observations from vessels or aircraft impossible. And then there are the sudden winds and storms. Perhaps in the future new technology, radio tagging and satellite tracking will help us to travel with and eavesdrop on the whales without ever leaving home.

FACING PAGE: *Arctic whales have no back or dorsal fin. Small front flippers and tail flukes reduce the amount of skin exposed to the cold water and therefore reduce heat loss.*

ASIA
RUSSIA
Kara Sea
Barents Sea
Sea of Okhotsk
Laptev Sea
ARCTIC OCEAN
Spitsbergen
Chukchi Sea
Anadyrskij
Bering Srait
Point Hope
Point Barrow
BERING
SEA
Norton Sound
Alaska
Queen Elizabeth
Islands
GREENLAND
Thule
Beaufort Sea
Lancaster Sound
Pribilof
Islands
Kuskokwim River
MacKenzie
Bay
Baffin Bay
Bristol Bay
Baffin
Island
Davis Strait
Repulse
Bay
Gulf of Alaska
MacKenzie River
Lab
Sea
Hudson Strait
Hudson Bay
CANADA

DISTRIBUTION AND NUMBERS OF ARCTIC WHALES

Bowhead Whale (Family Balaenidae)

Even though on a map of the Arctic seas it looks as if the whales could swim anywhere they wanted in the northern ocean, groups of bowhead whales are separated from each other by the permanent polar cap, land masses and ice. What the map doesn't show is the sea ice in the Northwest Passage preventing the Western Arctic bowheads from travelling across the central Arctic to join the Eastern Arctic bowheads on the Atlantic side. Scientists describe four or five groups or stocks of bowhead whales according to their geographic location. The greatest number of bowheads, about 7500, are found in the Western Arctic (known as the Bering/Chukchi–Beaufort Sea stock). There are only about 250 Eastern Arctic bowheads (known as the Davis Strait–Baffin Bay stock).

Bowheads move north in the spring when leads and deteriorating ice allow passage up to about 75° north latitude. After a summer feeding in open water, they migrate south before the winter freeze-up to 55° north latitude for the winter.

Narwhal (Family Monodontidae)

The majority of narwhals spend the winter in northern Davis Strait and southern Baffin Bay. In late June and early July, 15,000 narwhals will head for Lancaster Sound in a matter of days in search of the deep bays and fiords of northern Baffin Island. Some travel so far north that they are less than 320 km (200 miles) from the North Pole. An estimated 4000 narwhals head north and east to summer on the coast of northwest Greenland near Thule. A much smaller number, about 1200, pass the winter in Hudson Strait, then in the spring move into Hudson Bay to spend the summer in the northwestern section of this huge bay. Total world population of narwhals is estimated at 25,000.

Beluga (Family Monodontidae)

The beluga is by far the most widely distributed and the most numerous of the three species of Arctic whales. More than 60,000, and perhaps as many as 100,000, live in Arctic and subarctic waters adjacent to Canada, Alaska, the former Soviet Union, Norway and Greenland. A now-isolated population of 500 animals lives in the St. Lawrence River. At one time they were probably the southernmost of the belugas that ranged in one continuous sweep from the St. Lawrence River, north along the Labrador coast to Ungava Bay in Hudson Strait. Like the other Arctic whales (except for the St. Lawrence belugas), they spend winters in areas of open water or shifting ice, where they have access to air. In springtime they move north, sometimes thousands of kilometres, to a favoured summering area.

Extraordinary cold, killing winds, a short summer and lack of surface moisture make it impossible for all but the smallest and hardiest cold-adapted plants to live in the Arctic landscape. ROY TANAMI/URSUS

Chapter 2 SPRING

The wind-ruffled ocean glitters in the subarctic spring sunshine. In the blue air above, sea birds shriek and swoop on invisible pendulums. Occasionally chunks of white ice move in the distance, the first early-season icebergs shed from one of the more than one hundred fresh-water glaciers of West Greenland.

In the next four months, up to forty thousand icebergs will fall free of their mother ice rivers. Once free and in the ocean, they will circle counterclockwise in Baffin Bay. After travelling north along the coast of Greenland, they will arch left and move south along Baffin Island into Davis Strait. Large ones that haven't melted on the journey might be picked up by the Labrador Current and swept farther south, down the coast of Labrador, past Newfoundland and the Grand Banks. Such was the origin and movement of the iceberg that collided with the British luxury liner the *Titanic* in mid-April of 1912.

SPRINGTIME COURTSHIP

It is another April. Belugas and narwhals are mating in the offshore pack ice of Baffin Bay and Davis Strait, where they have spent the winter among the ever-moving sheets of ice. Although the water is cold, about –1.7°C (38°F), the whales' ardour is not. A large creamy white male beluga in vigorous pursuit of a snow-white female squeals his enthusiasm as they stream through the water. Suddenly the male throws down his tail, bends violently and

FACING PAGE: *Winter nights of twenty-four-hour darkness finally give way to spring sunshine and longer days, melting the winter ice cover and so unlocking the Arctic seas for fish-eating birds, seals and whales.*

swings his body around to confront another pursuing male. Bubbles pour from his blowhole as he throws his head up and down, shrieking his displeasure at the intruder and repeatedly clapping his jaws together with loud snaps. His eyes are wide and red rimmed, and the fat and skin on his forehead, or melon, vibrates violently. The noise is incredible, and the message is clear. The female slows, turns her head slightly, takes in the scene at a glance and continues to swim, propelled by graceful strokes of her tail flukes. She is not at all concerned.

When the big male is satisfied that his aspiring rival has been subdued, he sprints after the disappearing female with powerful pumps of his tail. His pectoral fins angle with exquisite precision, shooting his body in an arc to the surface for a breath of air, and then he is beside her again, swimming in perfect synchrony with her, stroke for stroke, breath for breath. He rolls onto his side to look at her. She turns her head, then her body, and with calm grace swims beneath him, her back barely caressing his belly as she crosses below him. And so the love dance begins.

For hours they swim, rubbing their bodies against each other in every conceivable way. Sometimes they are silent; sometimes one or the other trills and sings the strange language of belugas. They rub and roll and twine together, caressing each other completely without benefit of arms or hands, in a sinuous, sensual water ballet. Eventually the male begins to mouth the female—her flippers, her flukes and the base of her tail, first gently, then with more energy. Finally, she rolls beneath him, thrusting her lower belly against his underside. As he circles back around her, his erect pink penis emerges from a neat slit on his belly, and as he slides up beside her, his belly to hers, his penis disappears into her vent as they glide in perfect unison.

For whales of all kinds, mating is an act of absolute mutual consent. The degree of coordination required to successfully copulate in water with no arms and nothing to hang onto requires that it be so.

And so the whales pass the early spring in their winter habitat, waiting for the icy black claws of winter to loosen their grip on the high Arctic. Long hours of daylight through the spring and early summer will raise the air temperature, increase solar radiation and unlock rivers to pour warm water into the cold Arctic seas. As the sea ice deteriorates, winds will stir up the sea, hastening the melting of broken ice, and at last it will be the "time of open water."

MIGRATION

Arctic whales leave their wintering grounds south of the land fast ice and gather at the ice edge in early spring, watching and listening for signs of a crack or lead in the deteriorating ice. It is migration time. Time to move north.

In late April, Western Arctic bowheads move north undeterred by the ice cover that prevents belugas from starting their northern migration. These black giants literally break their own trail through the ice. Perhaps by listening for a difference in the echoes from their own sounds as they bounce off the underside of the ice, bowheads can distinguish between thick ice built up over a number of years, known as multi-year ice, and thinner, new ice that is only one winter old. The smooth, flat texture of new ice may reflect sound differently from rough, multi-year ice, providing the whales with valuable information about the ice thickness. Or perhaps they visually determine ice thickness by the brightness of light penetrating through the varying densities.

However it is that they know where to break through, they use their great weight and forward momentum to rise from below through ice up to 22 cm (8 to 9 inches) or more, absorbing the impact on a thick pad of fibrous connective tissue on the top of their heads. Having created a breathing hole, they replenish their oxygen supply with six or more powerful blows before moving on, travelling 3 to 6 km/h (2 to 4 miles per hour) for long distances under the ice before breaking through the ice to breathe once again. Each breathing place leaves a telltale hummock of ragged ice about a metre square (10 square feet) to mark the whales' determined passage north.

The underwater highway is strewn with hazards—shallows, underwater ridges and, at this time of year, massive ice floes trapped in surface ice. Such is the bowhead's navigational skill that it identifies underwater ice masses from a distance of a kilometre (about half a mile), alters course and swims around the floe, which may hang more than 40 m (131 feet) from the surface.

Over the thousands of hazardous kilometres, bowheads generally travel in the company of other whales, filling the sea with their low-frequency moans and screeches. Fifteen whales travelling together might spread out over 10 to 20 km^2 (4 to 8 square miles), maintaining constant contact through calls and countercalls, signature calls and mimicking sounds. The whales get more than simple companionship out of travelling together; they benefit from the collective experience of the group. Fifteen whales in constant auditory contact looking for a crack or lead through the ice that all can use, or finding new ice to break through, is more efficient than an individual attempting the trip alone.

Cracks known as leads open in the shore fast ice to create highways for the Arctic whales' spring migration.

After the bowheads, sometimes literally on their tails, come the belugas. When the ice cover is much more than 7 to 8 cm (about 3 inches) and too thick for them to break through, they follow the bowhead like a supply ship follows an icebreaker, using its path and breathing holes. Otherwise, belugas moving north from the Bering Sea towards Bering Strait and the entrance into the Chukchi and Beaufort Seas navigate the watery maze of ragged narrow leads on their predictable spring journey north.

Shore fast ice doesn't gradually melt and disappear; it fractures, ruptures and tears, opening deep blue veins in the white skin of ice. Some leads are so broad that dozens of animals can swim abreast. Others turn a sudden jagged corner, constricting the whales and forcing them to swim head to tail. When leads narrow or holes in deteriorating ice sheets confine the whales to predictable surfacing areas, patrolling bears watch for an opportunity to hunt.

FACING PAGE: *A thick, cushion-like pad on the top of this bowhead's massive head is sometimes used to ram ice cover from below to create a breathing hole during the spring journey north.*

DANGERS

The huge white bear lies absolutely still on the edge of the ice. It does not blink as it stares deeper and deeper into the inky water. Like scanners, the bear's eyes sweep back and forth, up and down, piercing the liquid darkness. Suddenly they lock onto a grey shadow. Still the bear does not move as it takes in the ghostly forms swimming their slow, circling ballet. As the bear watches, four or five belugas become a dozen, then twenty or more as they mill below in graceful random ovals. Large white adults are the easiest to see, but the bear is not looking for them. It is watching for smaller, pearly grey immature whales. With languid pushes of their tail flukes, the whales casually rise, approaching the surface. The bear knows that soon each of the whales will break the surface, exhale and inhale in one brief snort, and dive again.

Intense concentration vibrates in every white hair on the bear's body. Like a marksman with his finger on a hair trigger, waiting for the perfect shot, every cell in the bear's body is focussed on the perfect moment to strike.

Crash! The bear's huge curved claws drive through the water with vicious power and lightning speed, claws slashing the butter-soft hide of a young whale that is just about to breathe. Sensing rather seeing the danger, the whale flinches sideways in a flash of scaring pain. Terrified, the whale dives, and with each panic-driven pump of its tail flukes, black clouds of blood emanate in smokelike puffs from the hideous parallel gashes of ripped flesh. Eventually the bleeding will stop, leaving garish red parallel marks across the whale's back. In time the wounds will heal, but even as an adult the whale will wear a permanent badge of its brush with death in five long, chalk-white scars on its creamy skin.

The bear meant to scoop the whale from the water in one mighty sweep of its forearm but struck a fraction of a second too soon and missed. The rest of the whales evaporated as the panic of the one was telegraphed silently and instantly to the entire pod. Now the bear shifts its bulk, licks his bloodied paw clean and settles down to wait for another whale.

Sometimes on the northward migration, wind and cold conspire to close a lead and so trap some whales in a small pond of open water in a desert of solid ice before wind and warmth open the pathway again. Anxious and nervous, the whales circle, surface to breathe and dive, circle, surface and dive, sometimes for days on end. Although the ocean beneath them spreads forever in all directions, they cannot leave the breathing hole. This is when both the big white northern bear and the traditional hunter get close to the whales and when the whales have no escape. Bears and hunters watch for clouds of condensing air as

FACING PAGE: *The great white bear of the North sees the deteriorating ice and prepares to hunt the migrating whales at the floe edge.*

the whales expel litres of warm breath into the cold air. They listen too for the sound of breathing whales.

The name given to such an event is the Anglicized Greenland Inuit term *savssat*, meaning "whales or seals locked in a hole in the ice." The singular, *savsaaq*, refers to one entrapped animal. In Alaska the equivalent words used by Inuit from Bering Strait to Point Hope are *sapraq* (singular) and *saprat* (plural). For the *savssat* to be a hunting bonanza for bear or Inuit, they must remain trapped long enough to be detected. Although such areas are unpredictable, traditional hunters have long identified the areas where *savssat* are likely to occur and have developed certain travel patterns and settlements to take advantage of these locations.

The bear hunts trapped beluga or narwhal the same way it kills its staple food, the ringed seal (*Phoca hispida*) and bearded seal (*Eignathus barbatus*) by lying absolutely still on the edge of the ice with outstretched paws. As a seal or whale surfaces, in one powerful motion the bear stuns or kills it with a crushing blow to its head and at the same time pulls it from the water. Rewards of a successful kill are substantial, considering that a 28-kg (62-pound) ringed seal provides six days of food for a 230-kg (500-pound) bear, whereas a 600-kg (1300-pound) beluga offers the equivalent of 140 days of nourishment. In one documented case, a large bear (possibly a male), a female and three one-year-old cubs were seen near the carcass of a 2.3-m (7.5-foot) beluga hauled onto the ice. With the exception of the tail flukes, it was stripped of skin and blubber. Its front flippers and some ribs were gone, but its internal organs were intact. In another remarkable observation from an area south of Bering Strait at the end of April, there is evidence that good news travels fast. On first observation, fifteen bears and at least forty whale carcasses, primarily grey subadult animals, were seen. Two weeks later, twice as many bears were scavenging at the same site.

Savssat occur throughout the North American and Eurasian Arctic while the whales are travelling north during the spring and south during the fall. This phenomenon is part of the great power and unpredictability of Arctic winds, ice and temperature, and for all the Arctic whales, including the bowhead, it is a far more serious hazard than predators. Once the animals are trapped, if the cold persists and the hole begins to shrink, the outside edges freeze ever closer until the open space allows only one animal at a time to breathe. Eventually there is no opening left. All the whales die, not just the youngsters, from lack of air.

The hazards of travel are great, particularly for the young, who lack experience, are small and have less stamina than older whales. But accidents, foul weather, lack of food, long distances to travel, lack of breathing holes and predators are as much a part of the twice yearly migration as is the bounty that awaits the Arctic whales when they finally reach their northern destination.

FACING PAGE: *For centuries migrating whales confined to leads provided human hunters with good opportunities for whaling. Here photographers "shoot" the mighty bowhead as its huge tail flukes disappear near the ice edge.*

Winds and currents that churn and shift the floating ice to block a lead or close a breathing hole are as much a threat to the migrating whales as is the stalking polar bear.

Chapter 3 SUMMER

AN AMAZING TOOTH

On this summer day, the water's surface in Repulse Bay is a softly moving mirror to the blue sky, the thin white clouds and a few black silhouettes of birds. Windless and quiet, the entire surface of the sea is exposed. Suddenly the glassy plain is pierced from below by a brown-white horn, a long, spiralled pole pushed skyward by a narwhal, the swimming unicorn, one of twelve hundred narwhals summering in the area. Now another, then another tusk appears, as adult males leisurely loll in the water, crossing their tusks.

The tusk is a marvellous, mysterious strange tooth that emerges through the upper lip, slightly off centre to the left, and spirals counterclockwise for up to 2 m (6.5 feet) or more. Unlike the elephant or the walrus, the narwhal typically has only one tusk, not a pair; nor is there a hint of a curve in this straight-as-an-arrow marvel of dentition.

Great minds and even livelier imaginations have speculated on the function of this wondrous tooth. Some suggest that it is used to pick up sea grass, which is purported to be a favoured food, though this preference has not been scientifically supported. Others suggest that it is used to spear flatfish off the sea floor. How the whale then removes the speared fish and puts it into its mouth has not been explained. Or perhaps the narwhal male uses its tusk to stir up bottom sediments and flush out flatfish hiding in the sandy mud. Yet others propose that the purpose of the tusk is to break ice, to function as a wave guide for underwater vocalizations or to serve as a prop so that the whale can sleep at the surface, its tusk resting on an ice floe. Still others suggest that the tusk is a cooling mechanism to rid an

FACING PAGE: *Only adult male narwhals develop the bizarre spiralled tusk, which is in reality an upper front tooth that grows straight through the upper lip. Rarely, a double-tusked individual is found, and occasionally, a tusked female.*

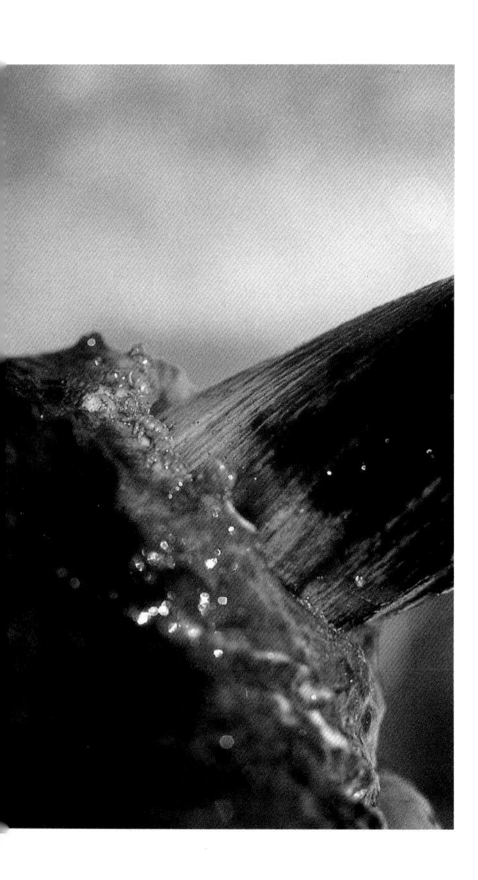

ABOVE: *Invariably the base of the narwhal's tusk is infested with lice.*

FACING PAGE: *On the summering grounds, adult males are often seen leisurely crossing tusks. Scientists speculate that the activity, known as "tusking," may be related to dominance or the maintenance of a social order in male groups.*

active whale of excess heat, since the tooth has a pulp cavity filled with nerves and blood vessels from base to tip. The reason that so little is known about the animal that owns the fabulous tusk while volumes have been written about the tusk's mythical origins, medicinal properties, restorative powers and monetary value is that the living animal has been observed by scientists for less than a generation, whereas the narwhal's tusk entered trade centuries ago.

With the exception of a rare tusked female, tusks are the property of males. The first evidence of a tusk appears when the male is a year old. Smaller than a finger and slightly larger in diameter than a pencil, it grows slowly for the first few years, and then, as the male narwhal achieves sexual maturity at age eight or nine, the tusk length and girth increase rapidly until the tusk is perhaps 2 m (6.5 feet) or more in length with a circumference of 23 to 25 cm (9 to 10 inches) and a weight of 6 to 9 kg (13 to 20 pounds).

The tusk is known as a secondary sex characteristic, an obvious sign of maleness and maturity, much like the beard on a human male. To what degree the size and length of the tusk command dominance or power in the social hierarchy is not known, but males spend a great deal of time lifting and crossing their tusks with each other. This activity, which is commonly observed on the summer feeding grounds and may be a year-round activity, has been dubbed "tusking." Whether this is a kind of muscle flexing to keep in shape for more serious duels that occur during the very early spring mating time is also not known, but at least one third of the males have broken tusks, suggesting that at some time in the adult male's life tusks are wielded in a less gentlemanly manner than is seen during the summer. Many old males wear long white scars on their foreheads, and one male was found with a broken tusk tip buried in his head.

Although the tusk base is invariably infested with whale lice and the shaft is stained greenish brown, the tusk end is polished to a gleaming white, indicating that males do something with their tusks that incidentally results in polishing.

BIRTH OF A WHALE

Eighteen hundred kilometres (1100 miles) to the south of Repulse Bay in the high Arctic, thousands of belugas have passed through Hudson Strait and into Hudson Bay for the summer. Here, and throughout the range of the Arctic whales, adult females have separated themselves from the migration herds. Some have dependent calves born the previous summer. Others prepare to give birth.

At noon one day in the second week of July, five female belugas swim leisurely into the quiet waters of a shallow inlet in eastern Hudson Bay. They have travelled for many days and are closing in on their final destination, where a northern Quebec river spills into Hudson Bay. There are two large white whales and three slightly smaller, pale grey ones. One female has a distinctly different shape from the rest; she looks like a big white balloon.

When the group stops to rest, four animals drift quietly at the surface while the swollen female takes a breath and sinks slowly to the shallow, silty floor, where she rolls over to rest, belly up, on the bottom. Her eyes close and she sleeps, holding her breath for twelve minutes. Wakened by the need to breathe, she floats to the surface, takes a few breaths and sinks to sleep again. The pattern continues for hours. The whale's languid movement and quiet sleep is occasionally interrupted by a brief, involuntary spasm, in reaction to quick pain.

She hasn't eaten anything today and shows no sign of hunger, though the other whales have long since wakened, dispersed to hunt and returned. Towards early evening she rises to the surface and moves a short distance away from the others at rest on the surface, but this time she remains at the surface. The top of her head, her blowhole and the forward half of her white back barely show above the dark water; the back half of her body is submerged. She opens her mouth, yawning under water.

For more than an hour, she drifts and occasionally opens her mouth. She is very calm, not straining or flexing, though she slowly moves her tail up and down. Then, a pale yellow cloud appears under her as litres of amniotic fluid flow into the water. At last she begins to swim at the surface, slowly but strongly, in large, easy circles. The birth process has started.

Something smooth, half-round and dark appears between the two longitudinal mounds of her mammaries on the flawless white expanse of her underbelly. The mysterious ellipse expands as each quarter hour slips past. Suddenly the "ellipse" protrudes—it is the top of the baby whale's head. According to traditional thinking among those who study whales, this is abnormal. Whales are supposed to be born tail first so that the infant can maintain its oxygen supply via the mother's umbilical cord until the last possible moment. In addition, if

FACING PAGE, TOP: *During the head-first birth of this beluga, milk jetted from the mother whale's mammaries with each contraction.* JEREMY FITZ-GIBBON

FACING PAGE, BOTTOM: *Within minutes the calf was delivered to just past its eyes. The top of the calf's head and its tightly closed blowhole are clearly visible.* JEREMY FITZ-GIBBON

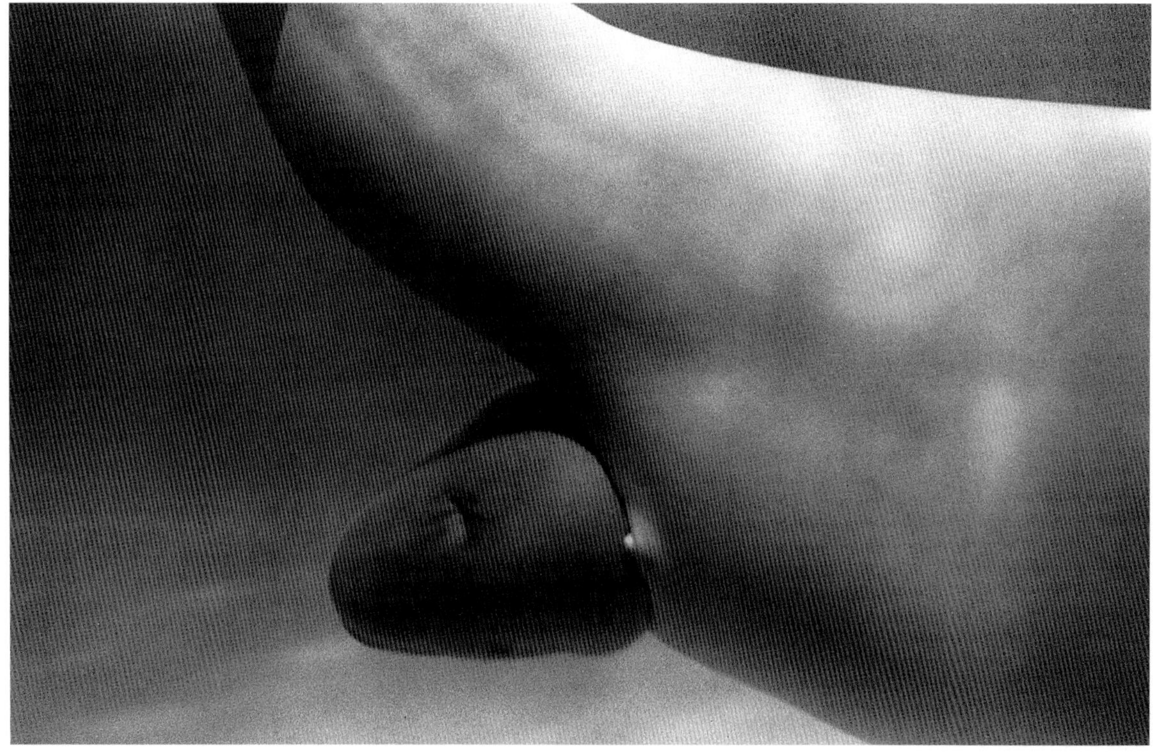

a baby whale's blowhole is exposed before it is born, it might try to breathe under water and drown, as any other air-breathing mammal would. The theory continues that when the calf is born and breaks free of its mother, the mother whale turns around, lifts her newborn on her nose and takes it to the surface for its first breath.

This birth is different. This whale is being born head first. With each contraction, the calf's head moves out a few centimetres only to be drawn back a few moments later. With each contraction, white clouds of expelled milk pour a gauzy screen over the scene. Ever so slightly the pace quickens. Stronger tail strokes move the mother a bit faster. Now the calf's entire head appears. Out to the eyes. Back. Out to the blowhole. Back. Up to the pectoral fins. This time it does not withdraw but hangs lifeless, but for a flaccid jiggle with each stroke of its mother's tail. Its eyes are open, round, black and blank. The mother continues to circle, with the peculiar dark grey appendage dangling half out of her body. Then, as if some magic switch has been thrown, the half-born calf opens its small, pink and toothless mouth and wags its head. It's alive!

Until now the pregnant whale's companions have left her alone. One or another of them have drifted over to her during the past three hours, then drifted back to join the larger group. Suddenly the four companion whales freeze. From out of nowhere appears a large white male. He is probably just passing by and is naturally curious, but now he is attracted by the alert behaviour of the females. The females, clearly nervous at the male's presence, vocalize in short, sharp blasts.

A few body lengths away, the mother whale pumps her tail flukes hard, and with one violent thrust the calf swims free in a burst of blood and fluid. Cold stings the little body like an electric shock. With limp little flukes, it first lurches towards the sea bottom, then throws its small, blunt head to the light overhead and swims vigorously to the surface. About one quarter of the calf's body pops straight out of the water before the whale flops over and drops back underwater without taking a breath. It is perhaps thirty seconds since he was born, and he has yet to take his first breath. The calf swims wildly, with no control, and again throws himself at the surface. Still he doesn't breathe.

Where all had been quiet and serene, there is an explosion of sound and activity. White whales, water and foam mix in one seething mass on the surface. Excited by the commotion, the big male tries to get at the calf. The water boils as frantic mother, newborn calf, aggressive male and attending females converge. The seconds pass, and the whales' vocalizations increase in volume and intensity. Is the mother attempting to establish a sound bond with her newborn? Is this a territorial response from the male? Is it fear, aggression? Under water the noise is deafening, but above water the sound is muted enough that when the calf surfaces for the third time a lovely little "pop" can be heard as he takes his first breath.

What a way to enter the world!

For 14 ½ months the little whale has floated in his own warm ocean within his mother's body. Suddenly he swims into icy cold, horrendous noise and huge white rubbery things. Everywhere he moves he is blocked by the bulk of another whale as he bumps into this one or slides off that one. Finally, the big male hangs back and swims off, followed by three of the four females. Only the mother and a light grey, teenage female hover near the calf, which is now breathing at regular intervals. He caterpillars forward with great energy, then throws his skinny little neck and head straight from the water for a baby puff of air, in contrast to his mother's exquisitely graceful motion as she swims upward, breaks the surface, exhales, breathes in and submerges again.

Whereas his mother is snowy white, the calf is a uniform grey-brown. At slightly over one third his mother's length, he is a fraction of her weight at 34 to 45 kg (75 to 100 pounds). Whereas she is a full, graceful curve from smiling mouth to trim flukes, he is thin and wrinkled, with an odd lump between his blowhole and his neck. His pectoral fins and tail flukes are wavy and soft, and the midsection of his snaky body is cleft in deep fetal folds like the rings around an earthworm. Altogether, he looks like something that was taken out of a box and unfolded.

As the evening fades, the little whale swims and flops randomly at the surface while being gently herded by the two females until they can position themselves on either side of him and thus determine the direction of his swimming. Once between the two big females, the calf finds himself being drawn along in the current created by his companions as they swim strongly forward. This is the whale's version of carrying the baby. Soon all three are swimming and breathing in synchrony.

During the night when the mother stops to rest, her young assistant keeps the calf swimming close to the surface; without a blubber layer, the calf lacks buoyancy and sinks when it stops swimming. The devotion of this young nursemaid is truly remarkable. Only a day ago she was full of fun, playful, an unabashed flirt. With the arrival of this calf, she is responsibility personified. Who exactly is the mother's young helper? Whether she is an older sister to the newborn, a young sister of the mother or no relation at all is not known, but the presence of a supportive female, an auntie, helping a mother and her newborn is also well documented in some other whale and dolphin species.

It has been an exciting half day of life so far, and the little whale has used up an abundance of energy swimming and keeping warm in water only a few degrees above freezing with little blubber insulation. Now he is hungry.

Land mammals such as primates hold their young to their breasts to make it easy for them to suckle. Other mammals care for their young in a nest or den, where they can curl around their offspring and offer them easy access to teats. Zebras and deer stand over their young, their udders within easy reach. The mother whale has no arms to hold her little one and no nest in which to cuddle him. Her mammaries are long and flat, located well to the rear on either side of her vent so as not to interfere with her hydrodynamic form. Her nipples are tucked neatly into two tidy slots.

Hunger and instinct draw the young whale to his mother's body. Although he sees very well, his search is by touch. He presses his cheek and then his forehead against her body and begins to explore the vast expanse of skin that looks and feels like the white of a hard-boiled egg. Travelling over and around the maternal mass at random, he is interrupted by short trips to the surface to breathe every twenty seconds or so. He slides beneath her, and pressing with the top of his head, feels his way, shadowed by her bulk. Suddenly his head pushes against something warm and soft, a much different sensation from the cool firmness of the rest of her body. The calf has discovered the swollen mammaries, and some primal message tells him that he is closing in on the target.

He pushes, bunts his mother's breast like a hungry calf, nibbles and noses, turns, surfaces to breathe and returns to his search for a teat. Rapid pumps of his flukes communicate his eagerness as he pushes and presses until he finds the tip of a teat between the folds of a mammary slit.

Neither the mother nor the calf can afford long, leisurely nursing under water, where neither can breathe. Instead they have a rapid delivery system that works something like this. The beluga calf wedges its lower jaw into its mother's vent while its upper jaw pushes into the slot that houses the teat. With its mouth half open, the calf rolls its tongue like a soda straw around the teat. The tongue, scalloped on its edges, makes a good seal at the base of the teat. The pressure of the calf's lower jaw on the mother's vent acts like a tap, and when the calf presses it, milk pours out. The calf doesn't suck; it just holds on and swallows as fast as it can. When the calf lets go, the flow stops.

The young Arctic whale, whether it is a 4-m (13-foot) bowhead calf or a 1.5-m (5-foot) beluga or narwhal, needs to be fed in vast amounts because it a large baby as baby mammals

FACING PAGE: *The whale calf suckles briefly under water from one of two mammaries in front of the mother's tail flukes.*
DAVID ROELS/
VANCOUVER AQUARIUM

TOP: *A rare photograph of a bowhead calf.*

BOTTOM: *Narwhal calves like this one, shown next to its mother, are born slate-grey and without tusks. As the whale matures, it will develop the spotted colour of the adult and, if it is male, will sprout the characteristic unicorn tusk.*

go, and it needs to quickly lay down a thick layer of blubber for both warmth and buoyancy. Whales have no access to fresh drinking water and don't drink sea water; rather, they metabolize their fluid requirements from their food. The traditional mammalian approach is to produce large volumes of milk, but that approach requires large volumes of fluid. Instead of producing more milk to meet the baby's needs, the marine mammal produces better milk, in a highly concentrated form that delivers more kilojoules per litre. Belugas produce a "high-octane" milk with up to ten times more fat and five times more protein than an equivalent volume of cow's milk.

It doesn't take the new calf long to solve the food problem, though initially he tries to suckle from the young female as often as he does from his mother. Although only his mother is lactating, both females respond the same way. When he is ready to suckle, the calf falls back from his position between the females, swims underneath one and pushes the top of his head near the mammary. The female slows, holds her tail flukes steady and rolls slightly to one side as the calf latches on. As the female glides forward, the calf pumps his tail vigorously during the twelve or thirteen seconds of each feeding bout. The calf lets go, rises to the surface, breathes and either nurses again or settles down for a nap on the back of one of the females.

Within 2½ days a regular nursing pattern is established between mother and calf, and the young nanny withdraws her services as abruptly as she volunteered them. In a remarkable change of roles, she sheds her responsibilities like a teenager out of school and returns to her mischievous, fun-loving self, blowing bubble rings, spitting water and playing underwater throw and catch with sticks, rocks, shells or seaweed.

She'd like to play with the new calf too, but he sticks to his mother as if held by an invisible tether. Even when groups of mothers and young calves are together, the little whales never stray. If they did separate, how a mother whale would recognize her calf is a mystery. Other marine mammals, such as seals, recognize their young above water by scent. After days at sea feeding, a mother fur seal finds her own pup among hundreds of others on the rookery by its unique scent. But whales don't have noses to smell under water. Perhaps a youngster recognizes its mother's voice. In any case, the calf cannot afford to be separated from his mother, his source of food, protection and transportation. Without her, he could never reach the river estuary.

All three Arctic species are typically born in the late spring and early summer; the timing is likely related to when food is most abundant and easily available to the mother whale. A long nursing period—twenty months or more for belugas and narwhals and perhaps that long or even longer for the bowhead—means that for a period of time the young whales do not eat anything directly from the environment but are absolutely dependent on their mothers for all their nourishment.

SUMMER DISTRIBUTION

Arctic whales share common migration corridors on their way north for the Arctic summer, but once there, they disperse, the narwhals into deep northern fiords, the bowheads to plankton-rich areas and the belugas to cluster, sometimes by the hundreds, with great predictability, at river mouths in Alaska, the Western Arctic, Hudson Bay, the Eastern Arctic or northern rivers in the former Soviet Union.

Like the swallows returning to Capistrano, belugas return to their particular estuary with great persistence and fidelity. Belugas choose their river mouths, and some are not as popular as others. Females and their young use the estuaries in disproportionately large numbers, and they are the first to return after being frightened away by boats or hunting. The estuary is a special place for them.

Biologists propose a number of theories to explain the belugas' attraction to selected freshwater outflows. Some say that females come to river mouths to give birth because the water there is generally warmer than the surrounding sea by up to 10°C (18°F). Since newborn whales have little blubber, warmer water would be a definite advantage in the first few weeks. But many females arrive at the river with calves that were born earlier, in the pack ice. In addition, whales of all ages and both sexes visit the estuary, not just females with calves.

Some scientists believe that warmer water enhances circulation and blood flow to the whale's skin and helps the animal to grow new skin faster, an advantage during the moult. Whether all belugas actually moult a layer of skin annually is still up for debate.

Some observers believe that good feeding opportunities attract the whales to the rivers. But because stomachs of whales hunted in one estuary were examined and found empty, some scientists claim that the whales are not feeding there. In other rivers, belugas appeared to be eating whatever was available.

Hunters so successfully exploited the unfailing predictability of belugas to return to certain rivers that these whales were totally wiped out in the Great Whale and Little Whale Rivers in sixteen short years of hunting between 1854 and 1870. Today, even if hunting is limited to a few whales killed in an in-river hunt, the entire herd is driven from the estuary in the process. Although they return a few days later, another hunt drives them out again. How many times can they be displaced from their habitat without some serious, but invisible, damage to the survival of the young and vulnerable?

Why some, perhaps all, belugas return to the same estuary or series of estuaries year after year is likely a combination of factors. They know what these factors are, but we don't. For the whales' sake, it is essential that we learn what the estuary means to the belugas.

FACING PAGE: *Newborns stay with their mothers as if stuck to them with invisible magnets. It will be many, many months before a youngster ventures more than a few metres away from its mother.*

BOWHEADS IN WATERY PASTURES

Dull, chronic hunger dogs the half dozen bowhead whales as they fan out over 20 km (12.5 miles) of black sea through small green-black fish, iridescent comb jellies and tiny floating particles that catch the light like dust in a soft wind. The whales' strangely melodic calls of bongs and groans ring out against a background of never-ending ice noise and the occasional descending trill of ringed seals.

After a search of many hours, the whales find what they are looking for—a living soup that spreads rich and dense on the sea's surface. Within a few minutes, the scattered whales converge on the floating food and the feast begins. They are eating small, shrimplike animals called copepods and euphausiids, ranging from the size of a house fly to that of a large grasshopper.

Concentrations of the floating swarms might be here and not there, or here today and somewhere else a week from now. This is because the tiny animals that make up the bowhead's diet have their own annual cycles of movement, growth and reproduction, and their own habitat needs and nutrient requirements. Their cycles of reproduction and growth are governed by light and nutrients.

In a flower garden, as light levels increase and days lengthen in the spring, grass and weeds seem to sprout overnight. The same process takes place in springtime in the northern oceans. With longer days and more intense light, tiny floating marine plants take up nutrients dissolved in the water and bloom into a bountiful crop of marine algae (phytoplankton). The plants in turn provide food for tiny floating marine animals (zooplankton). Together this drifting mass makes up what is known as the plankton. Some species of these floating marine animals, known as krill, become the primary food for the bowhead whale. When water temperature, nutrient supply and light levels are just right, krill reproduces into the billions, grows rapidly and forms a dense swarm that can range from the size of a blanket to many kilometres in all directions. These pastures, alive and floating, are what draw the bowheads to the Arctic as soon as the breakup of shore fast ice unlocks the pantry door.

A dozen or more gigantic bowheads move in a classic V formation at the surface, mouths wide open as they skim surface krill. Other times they feed lower in the water column, staying down ten minutes or more, rising for a series of breaths, then descending again and again. Feeding lasts for hours or days, as long as the food lasts. Sometimes food is on the bottom, and muddy water streams from their mouths as they surface. And always, as the huge whales feed, gigantic clouds of faeces dissipate into the surrounding water.

FACING PAGE: *The bowhead's upper jaw rests like a lid on its massive white-trimmed lower jaw, visible here through the water. Note the paired blowholes on top of the whale's head.*

An adult bowhead moving slowly with its mouth open exposes an opening the size of a garage door. The whale has no teeth. Its mouth is a huge chamber hung with 3- to 4.5-m (10- to 15- foot)-long curtains down each side. The "curtains" are made up of narrow strips, like vertical blinds, of a material known as baleen, or whalebone. This highly elastic, horny material is fringed on its inner edge and smooth on its outer edge. Hundreds of individual baleen strips, or plates, hang a finger width apart on each side of the upper jaw, the inner fringed edge of each strip overlapping the strip next to it. Lower ends hang free and are kept from spreading outside the whale's mouth by great, arched, rubbery flaps like upside-down cheeks on the sides of the whale's lower jaw.

Tonnes of water, krill and bits of floating seaweed, an unwary sea bird and a few small fish pour into the enormous maw as the whale ploughs forward and sea water flows out through the baleen. As the watery soup passes over the fringed mat of the baleen, solid material is trapped in the baleen sieve. The mouth closes and geysers of water pour out the sides, an enormous tongue licks krill off the inside of the baleen, and the gigantic throat convulses as the whale swallows.

Six hundred baleen plates, three hundred on each side of the whale's mouth, make up the bowhead's fishing gear. Housing this apparatus takes space, and that is why the bowhead's head consumes fully one-third of its body and why the whale's upper jaw is bowed up and out to accommodate the great numbers and length of its baleen plates. Like human hair and nails and cattle's horns, baleen is non-living tissue except where it is rooted and grows, in the gums. Like our own fingernails, baleen continues to grow as the ends are worn away. Although nine other species of whale have baleen, none is as long or as dense as the bowhead's.

In its own search for food during the Arctic summer, the bowhead whale will cross paths with belugas in the west and both narwhals and belugas in the east as they too search for food. The three species hunt and eat different things for the most part, at least in the summertime. Virtually nothing is known about their winter feeding habits.

BELUGA BANQUET

Whereas the bowhead's mouth offers evidence of what the bowhead is eating, the beluga's forty sparse, gum-level, peglike teeth don't reveal much. It's not eating krill, or it would have baleen plates. Its teeth are too widely spaced for chewing and too flat and low to catch and hold prey.

With its flexible neck and supple body, a beluga can twist, turn and manoeuvre in confined spaces to corner prey, which it captures not with its teeth but by suction. A fish swimming a hand span from the beluga's mouth is sucked in as if snapped back on an invisible elastic. The whale selects, tests and accepts or rejects different food items this way. If a spiny crab, for example, is sucked in, tasted and rejected, it's blown out as fast as it went in. The crab, unchewed and still alive, can scurry for cover and live another day, or the whale might toy with it for a while in an undersea game of cat and mouse. When the whale is ready to swallow, its big pink tongue squeezes water out of its mouth to prevent it from swallowing sea water, which would add salt to the whale's diet, a problem for an animal that has no access to fresh drinking water.

The original size of an item of food doesn't matter to us. With knife and fork, we can reduce a cow, a pig or a chicken to bite-sized pieces. The beluga has no way of cutting a large food item into something more manageable, so everything it catches must be eaten whole or not at all. Consequently, a beluga's fish dinner cannot be more than about 4 kg (about 9 pounds). Gluttony can be a hazard, as demonstrated when a beluga was found choked to death on a 9-kg (20-pound) cod.

Belugas are opportunists, eating different fish and invertebrates according to season and availability. These include capelin, herring, salmon, saffron cod, Arctic char, flatfish, sculpins, shrimp, squid, marine worms and other muddy-bottom invertebrates. Arctic cod are an important food, especially in the fall, when the they congregate in huge schools and move into coastal waters for late autumn and winter spawning. The cod's popularity, not only with belugas, but with ringed seals, narwhals, sea birds and even other fish, is doubtless related to its abundance and extensive range, which extends to 84°42' north latitude, farther north than that of any other fish.

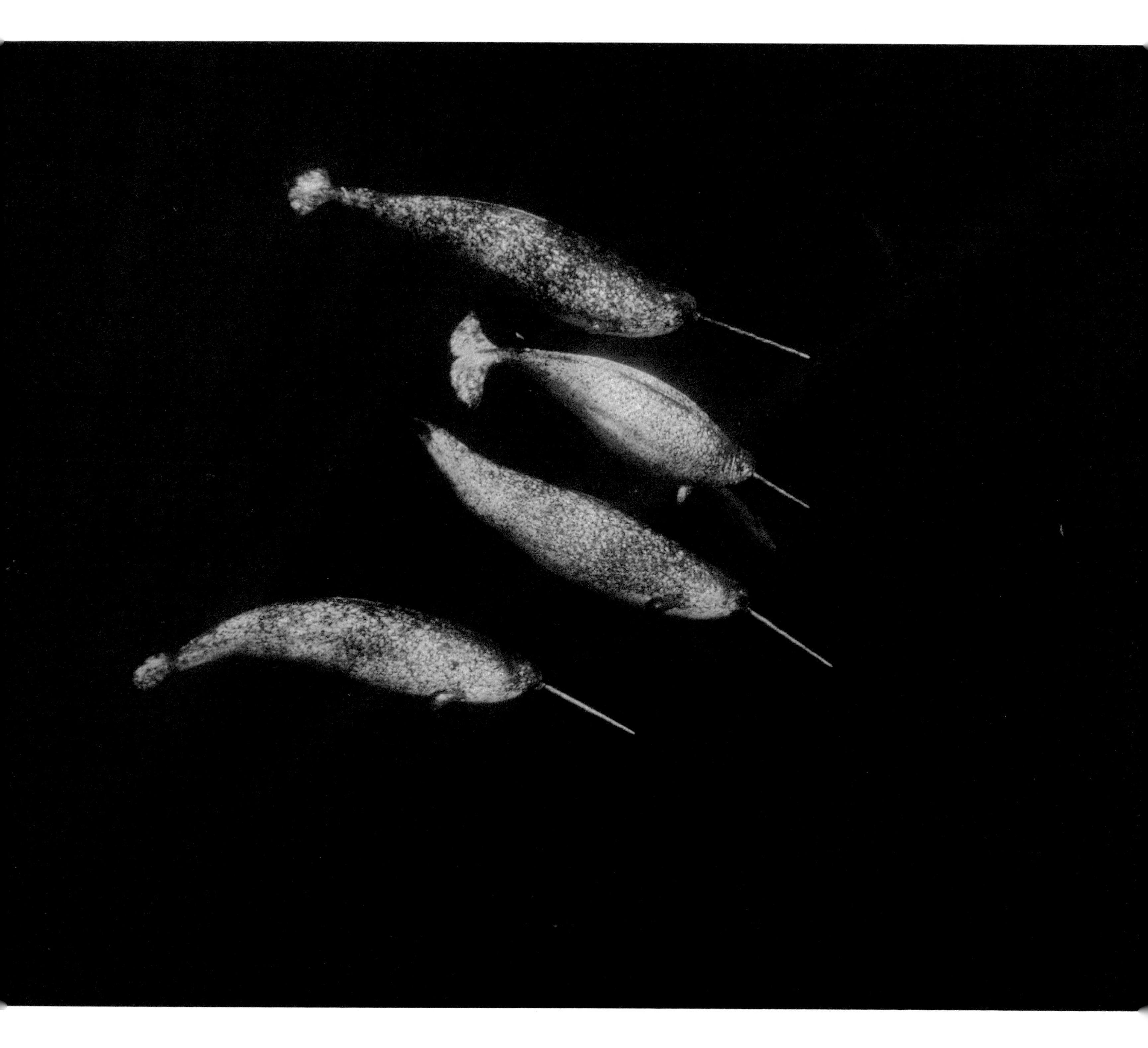

THE MYSTERY OF NARWHAL FEEDING

Like so much about the narwhal, its feeding behaviour is a mystery. The animal is well fleshed, even robust, yet it has a small mouth and no functional teeth, and the adult male hunts and eats with a tooth longer than a broom handle protruding from the front of his head.

Perhaps the narwhal uses the same suction method as the beluga, which typically looks over its food carefully, turning its head as it rolls from side to side to size up an item before it sucks it in. Deep-water flatfish found in stomach samples from male narwhals suggest they hunt in deep water, but how the narwhals see the fish at depths of 800 m (2625 feet), where there's not much light, is a mystery.

Perhaps the narwhal stuns its deep-water prey with a blast of sound, as has been suggested for the sperm whale, another deep-diving, virtually toothless whale that feeds on huge, deep-water squid. Narwhals eat squid too, and lots of it, as well as Arctic cod, Greenland halibut and a pelagic shrimp.

When scuba divers go under water, they take their own air supply with them, continuing to breathe as they dive. Their time under water is limited only by the volume of air in their tanks. When deep-diving whales, including the narwhal, dive, they have only their last lungful of air. They work hard to drive their bodies to the sea floor and search for and kill prey. Each underwater trip can easily be a kilometre or more.

Such a feat is made possible by this and other whales' marvellous adaptations for oxygen storage and conservation. Not only does a whale have twice as much blood volume as an equivalent-sized land mammal, but its blood cells hold ten times as much oxygen. Further, the whale is less sensitive to the buildup of carbon dioxide (this is what eventually forces us to breathe when we try to hold our breath), and its muscles are less sensitive to both oxygen depletion and lactic acid buildup, the thing that makes our muscles scream in agony when they are overexerted. In addition, the narwhal has a complex network of blood vessels between its lungs and spine to ensure that its brain and heart get essential oxygen while less vital organs wait.

Back at the surface, the narwhal exhales in an explosive burst, then sucks in huge lungfuls of air as it lies at the surface, its back heaving as it inhales. Minutes later, oxygen debt repaid, the whale is ready to go again. It swims slowly forward, picks up speed and, moving fast now, sucks in a last lungful as it arches its back, throws its flukes skyward and drives its body down into the mysterious black deep.

FACING PAGE: *Narwhals, the deepest divers of all three of the Arctic whales, search for their prey on the sea floor.* FLIP NICKLIN/ MINDEN PICTURES

SUMMERTIME PREDATOR

Belugas, bowheads and narwhals are drawn like magnets north to the Lancaster Sound area off northern Baffin Island by the abundance of food. Other marine mammals—ringed seals, bearded seals and walrus—go there to feed too. From the sound they disperse into adjacent fiords, channels, inlets and bays according to their particular needs, some to bear and nurse their young, all to forage and socialize. But life is not all a summer idyll for the Arctic whales. A predator follows in their wake.

At 8:00 P.M. in mid-August, two hundred narwhals are in Koluktoo Bay, one of the many deep fingers off Eclipse Sound and Milne Inlet near Lancaster Sound. Adults dive, youngsters play with bits of floating seaweed, big males leisurely cross tusks and, as at a cocktail party where everyone talks at once, whale sounds of clicking, squeaks and whistles are everywhere.

Then, unheard and unseen, a dozen silent black enemies slide into the bay. Three slick dorsal fins slice upward as a trio of enormous bulls breaks the surface in advance of the pod. A single, piercing killer whale call erases all sound in the bay. Now another and another reverberating call stabs the thick silence as the intruders signal to each other.

A large male narwhal brings his tusk down hard on the back of a nearby animal. The whales move nervously to the end of the bay. Clumped together they hover in silent terror.

Big and powerfully built predators, killer whales have such speed and stealth, along with cooperative hunting strategies, that nothing in the ocean, save large sharks, is safe from their attack. The killer whales cruise ominously across the bay, then turn in a wide arc, listening for the narwhals, waiting for them to make a sound and reveal their presence.

For most of the year, narwhals can evade this voracious enemy. In the ever-moving pack ice, or in leads through sheet ice or at the ice edge, narwhals flee to where killer whales are loath to follow, under the ice edge or into the pack ice. With their tall dorsal fins, killer whales are at a disadvantage in such places. But here in Koluktoo Bay, the narwhals have nowhere to hide, and so they move to shallow water. Inuit hunters recognize this behaviour and call it *ardlingayuk*, which means "fear of killer whales."

For a dozen minutes, the narwhals are completely silent. They lie at the surface, barely moving. Their strategy may be to reduce the total water volume they need to defend or, by clumping together, to take advantage of many eyes watching the enemy and gain anonymity in a large group. By keeping their backs to the shore, they needn't cover the rear, and by retreating to the shoreline, a noisy area under water, the narwhals may be able to confuse the echolocation and other sensory capabilities of the killer whale and avoid detection by the listening enemy. It could be any or all of the above, and whatever it is, it works for a while.

The killer whales haven't eaten for two days, and they are hungry. They wait, they watch, and they listen for an opportunity. Sometime in the brief Arctic summer night, a small group of narwhals moves across the bay. The sea explodes. Killer whales come from out of nowhere and at incredible speed, breaching full body lengths at the surface to overtake the frantic narwhals that in their panic flee in all directions. A young female narwhal only glimpses the 7-tonne (8-ton) bulk of a 7-m (23-foot) killer whale as it rams her midsection, driving her crushed ribs into her lungs. She is unconscious as the enemy whips around with a killing side slash of its powerful tail. Twelve killer whales attack the narwhal carcass. It shakes violently as the killer's forty-four conical teeth rip mouth-size chunks off what was minutes ago a whole living animal. Predator and prey boil in a bath of blood and foam. Soon all that remains of the kill is an oily slick on the water's surface.

Other narwhals escape with chunks of skin torn from head or flukes to expose ugly pink-white patches of raw blubber. If the victims survive, their wounds will leave permanent scars as a reminder of the attack.

There will be other victims before the short summer is over, as they fall prey to the estimated forty killer whales that come every August to hunt the marine mammals of Eclipse Sound. Each adult killer whale will consume, on average, the daily equivalent of about 68 kg (150 pounds). Not all victims will be narwhals. Some will be belugas, seals, fish or even a young bowhead whale.

Dozens of killer whales move into Lancaster Sound in the summertime, when the area is ice free and there is an abundance of prey—seals, belugas, narwhals and even young bowheads.

Chapter 4 FALL

A WORLD OF SOUND

A century ago, at the beginning of autumn, the lone sailor on watch aboard a whaling ship heard alien sounds blend with the rustling of the ship's great sails. The sounds came from the black water. Below deck, in wooden bunks built into the ship's groaning timbers, sleeping seamen were drawn from their dreams by deep, eerie, resonant sounds that filled the sea, penetrated the ship's planks and filled the waking sailors with its ghostly strangeness. Deep knocks, slow creaking like a huge door moving on rusty hinges, deep, sonorous moans. The sounds were strange, and strangely soothing. They had fullness and volume but no threat or aggression. These were the sounds of the great whales as they called to each other under water across kilometres of ocean, sounds that inspired stories of phantom ships and seductive mermaids luring sailors to their death. Today, if you go where the whales go, and if you are quiet and listen, you can still hear their calls.

Whales have little if any sense of smell and can't see in dark or turbid water, so they use sound to communicate the important messages that land animals relay to each other through vision and scent—territoriality, fear or a readiness to mate. By listening to returning echoes of their own sounds, whales also use sound to find food. And as they leave the high Arctic for their winter homes, they use sound to navigate.

Sound travels fast in air and four times faster in water—between 1450 and 1570 m (between 4757 and 5150 feet) per second, the speed increasing with temperature and salinity. It also travels farther in water than in air. Some scientists believe that large whales can

FACING PAGE: *Unlike bowheads and belugas, narwhals have been sighted only a dozen times during the last century in the icy waters off northeast Asia and Alaska.*

communicate across oceans—or rather that they could before propeller-driven ships generated sound in the same range.

Whales and dolphins makes sounds above water when they force air out of an almost-closed blowhole. The sound is similar to the sound produced when you stretch the neck of an overfilled balloon sideways to make the air whistle as it escapes. The whale does the same thing under water, emitting a trail of fine bubbles from its blowhole. Yet other sounds are produced inside the whale's head as it constricts and forces air between a series of nasal sacs between its blowhole and its larynx. Because there are no bubbles when this happens, it's difficult to know which whale is making the sounds when you are watching and listening at the same time to a group of whales under water.

Somehow cetaceans use the fatty tissue on their melons to direct a broad or narrow beam of sound of such intensity and quality that they can precisely analyze the returning echo for information. Some can use this technique to find fish less than a few centimetres long in the dark. With its eyes covered, one cetacean trained in a research experiment distinguished in seconds the difference between glass, plastic, aluminum and copper objects of exactly the same size on the basis of their acoustic properties. Scientists refer to the animals' use of these sounds as echolocation, echo-sensing, bio-sonar or sonar. Whatever it's called, it's a highly evolved sound and hearing ability that gives the whales both magic eyes to "see" with sound and invisible hands to "touch" at a distance using sound.

If the apparatus to send out the sound is sophisticated, then it follows that the receiving, or hearing, mechanism will be equally so. Whales have no ears that you can see. In some, a tiny pinhole is all that remains where a large auditory canal at the centre of a large ear lobe might have been on the whale's long-since-extinct terrestrial ancestor. Whales don't need ear lobes to gather airborne sound waves, since the inner ear is what does the hearing. Scientists believe that sound, in the form of returning echoes, travels through skin, fat, muscle and bones in the whale's head and then is transferred through windows of fat at the rear of each lower jaw and thence to the inner ear. Right and left inner ear mechanisms are isolated from the whale's skull by a cushion of bubbly foam. This separation ensures that the whale can compute the exact direction of the sound by the minuscule difference in time that it takes for the sound to reach one ear before the other.

FACING PAGE: *The fatty tissue on the whale's forehead (known as the melon) is believed to act as a lens to direct and focus underwater sounds. The whale's ability to use sound to navigate, communicate and "touch" at a distance is yet to be fully understood.* ROY TANAMI/URSUS

Arctic whales overcome special challenges of life in the ice by using sound. Studies of beluga echolocation show that belugas can accurately detect targets by sound even in the exceptionally noisy Arctic environment with its cacophony of sound and reverberation from weather, moving ice and underwater animal noise from millions of popping shrimp, walrus and seals. Belugas can detect sound bouncing back from the underside of the water's surface, and it is speculated that this is a special adaptation for under-ice travel and the need to detect air pockets, open water areas and breathing holes.

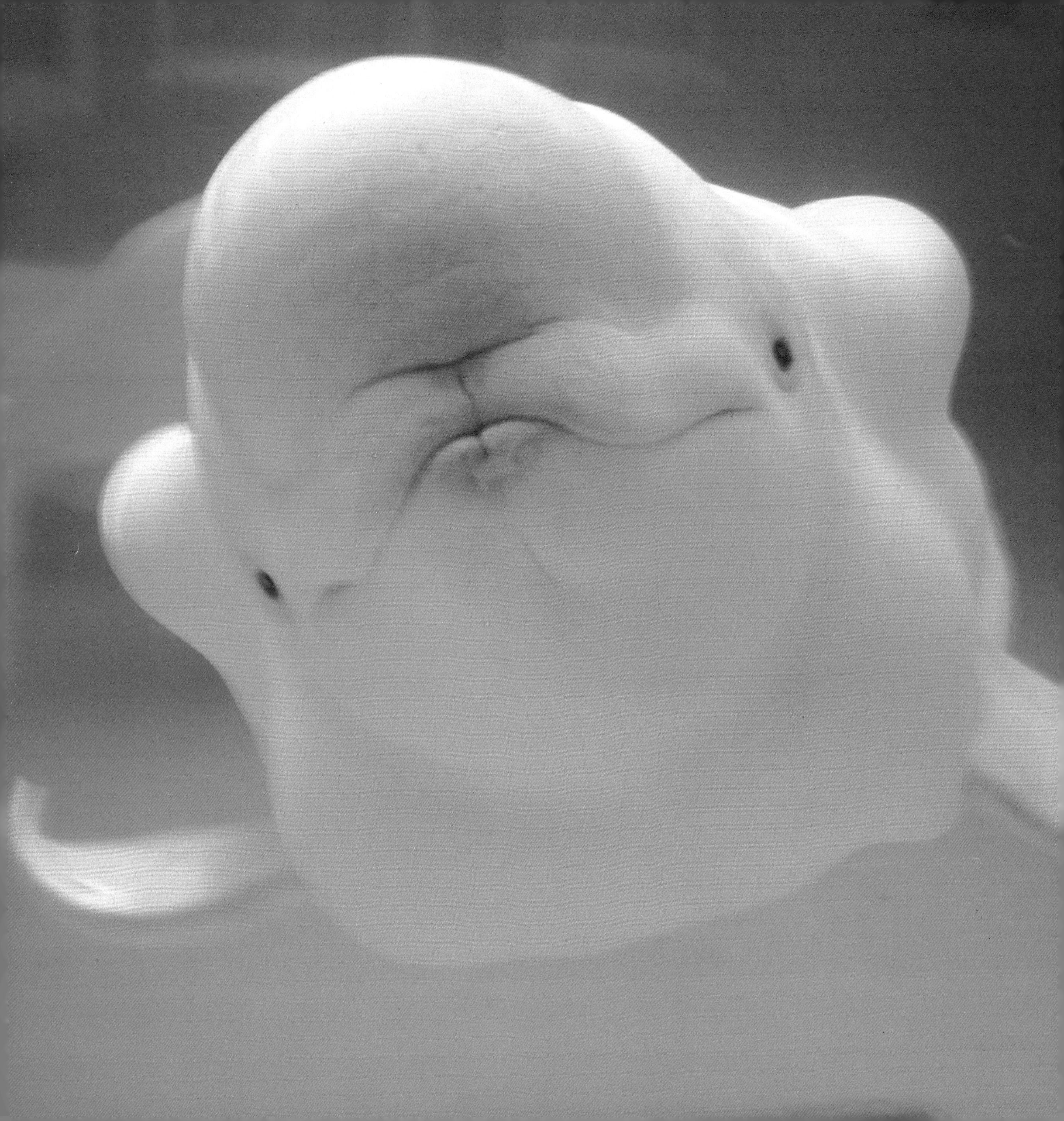

With their communication sounds, belugas make so much noise that whalers dubbed them "sea canaries." Male narwhals are noted for their large vocal repertoire when socializing, and even bowheads vocalize with enthusiasm when in the company of other bowheads. Why the Arctic whales are so chatty compared with temperate cetaceans might have something to do with who is or isn't listening. For most of the year, killer whales aren't around to hear. Tropical and temperate cetaceans are more vulnerable to predators and therefore might have good reason to be quieter and not attract attention to themselves.

Since different whale species in different environments produce and use sounds in different ways, to generalize is impossible. Some whales have been studied; most have not. Some make sounds in streams of a thousand individual clicks per second. Many make sounds above and below our hearing threshold. The whale's ability to use sound is so vast that we can hardly imagine what questions to ask, much less determine how to measure or test to find the answers.

SURFACE ACTIVITIES

Whales spend only 5 or 10 per cent of their time at the surface; otherwise, they are under water and out of sight. Nevertheless, the Arctic whales reveal something of their rich and varied social lives through their surface activities. Bowhead whales slap their tails, smack the surface with their huge pectoral fins and even propel their huge bulk skyward, one-third out of the water, landing on their chests with an astounding splash. This is sometimes repeated again and again, up to twenty times. One whale will hit another with a mighty slap of its flipper, or trios of whales will rise from the depths, flippers draped over each other as if hugging, then roll and caress in sexual play while sounding a marvellous combination of raucous trumpeting, moans, moos, upswings, shrill squeals, crunches and slaps. This might be serious courtship or the entertaining sex play characteristic of subadult cetaceans of many species. Suddenly there is a vigorous chase; then all is quiet.

The quietly cruising beluga barely breaks the surface to breathe its quiet little puff-breath, completely belying the energy and range of its social, sexual and play activity. Belugas together constantly vocalize and swim around, under and over each other. They seldom if ever breach but rather bounce vertically about a third of their body length above the surface. They play endlessly with objects, alone or with a companion, passing a piece of wood, plant material or even a dead fish back and forth. If there is no willing playmate, the whale will carry its play object around on its head. One beluga was sighted carrying a pair of antlers on its head, probably from a drowned caribou. Another whale will make an air bubble ball and suck it in and out as it follows the bubble to the surface. Or a whale will entertain itself blowing perfect bubble rings, like smoke rings, then watch them swirl and disappear just as the Arctic whales themselves disappear into another part of their watery world for the season of cold and darkness.

ANOTHER MIGRATION

Fall signals its approach with shorter, cooler days, and in response the Arctic whales prepare for their autumn migration before winter sea ice blocks their path. In the high Arctic, bowheads, narwhals and belugas begin leaving as early as mid-August. Others, such as the belugas summering in Hudson Bay, won't leave until early September.

Bowheads by the thousands in the Western Arctic and by the hundreds in the Eastern Arctic have feasted during the season without competition from any other species of baleen whale. All feed with greater intensity as the fall approaches, taking full advantage of the increased fat content of late-season krill. But by early September, bowheads in the Western Arctic leave the eastern Beaufort Sea and Amundsen Gulf and travel west, clearing Point Barrow to continue their westward movement across the Chukchi Sea, down the Asian coast and south through Bering Strait to their wintering grounds in the Gulf of Anadyr. In the Eastern Arctic, bowhead mothers and calves leave the high Arctic in August. Groups of subadults and large males leave the east coast of Baffin Island and the Isabella Bay area later. All move at a leisurely 3 to 6 km/h (about 2 to 4 miles per hour) to wintering grounds off southeast Baffin Island or on the Greenland side of Davis Strait.

Like the bowheads of the Western Arctic, belugas also leave the Beaufort Sea, travelling approximately 1200 to 1500 km (750 to 950 miles) to the Bering Sea. They travel faster than bowheads at 9 to 10 km/h (about 5 to 6 miles per hour) or more. Belugas summering in the high Arctic and the Eastern Arctic in areas such as Cumberland Sound, southeast Baffin Island, Hudson Bay and Ungava Bay move to Hudson Strait, Davis Strait and Baffin Bay for the winter. The move is necessary because the summering grounds freeze over in winter. Narwhals too leave northwest Hudson Bay and the high Arctic inlets and fiords for the close pack ice of Davis Strait and Baffin Bay, travelling in separate groups of adult males, subadults, and females and calves in groups typically ranging from two to twelve animals. Beluga groups seem to follow the same general pattern. Sometimes groups come together in loose aggregations to form a herd. Although it is clear that all the Arctic whales are highly social, the membership and relationship between the group (or pod) members is not clear. Mothers with young calves deliberately avoid all-male groups in the summer. Do they continue to do this on the wintering grounds? Subadults are seen grouped together, but at what point do they leave their mothers to live independently? Or do daughters remain with their mothers in female groups so that the subadult groups are all males? For the most part, these questions remain unanswered.

FACING PAGE: *Far fewer bowheads today than less than a century ago make the annual fall migration. Scientists estimate that 60,000 swam in Arctic waters before the days of commercial whaling. Today there are perhaps 10,000, only 15 per cent of the original number, seventy-five years after commercial whaling ceased.*

ABOVE: *King and common eiders are spring and summer residents of the Arctic but must leave in fall to overwinter in warmer climes.*

PAGES 66–67: *Because whales spend only 5 to 10 per cent of their time at the surface, much of their behaviour and many of their social activities are hidden from our view. Perhaps one day technology will allow us to observe the whales without disturbing them in their natural habitat.*

PAGE 68: Beluga Hunters, *a print by Rex Kangoak.* COURTESY OF INDIAN AND NORTHERN AFFAIRS CANADA. REPRODUCED BY PERMISSION OF THE ARTIST.

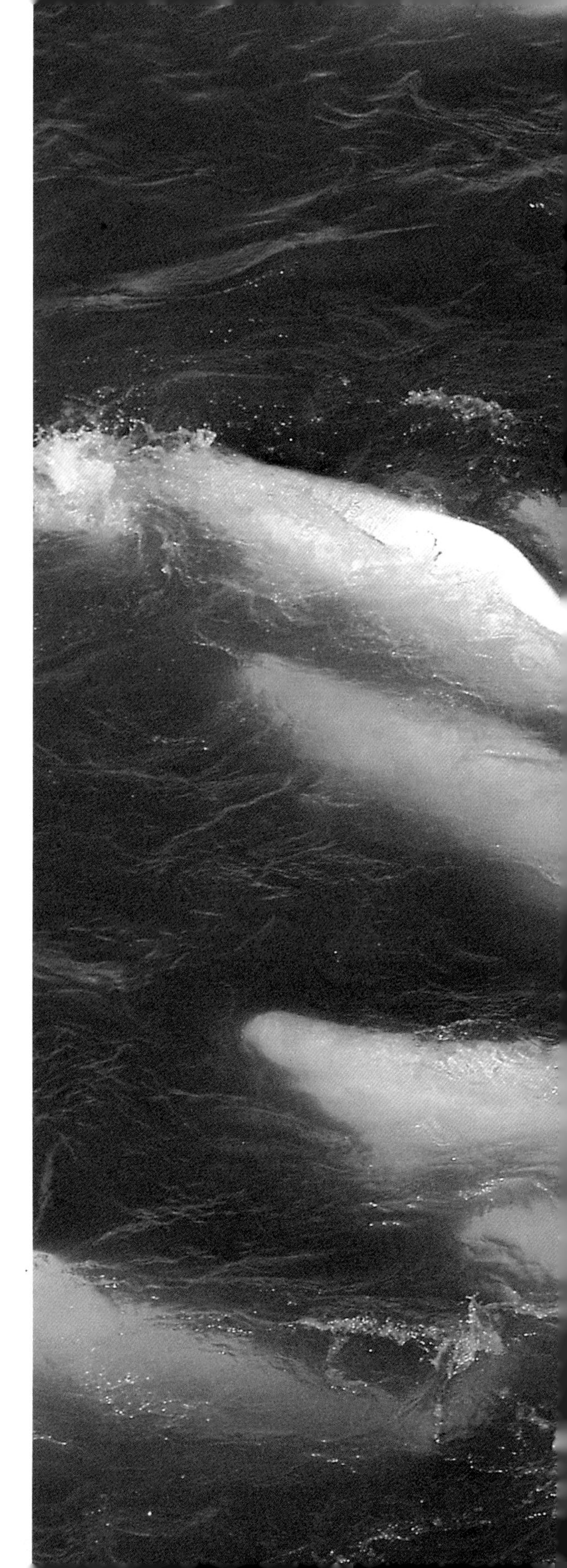

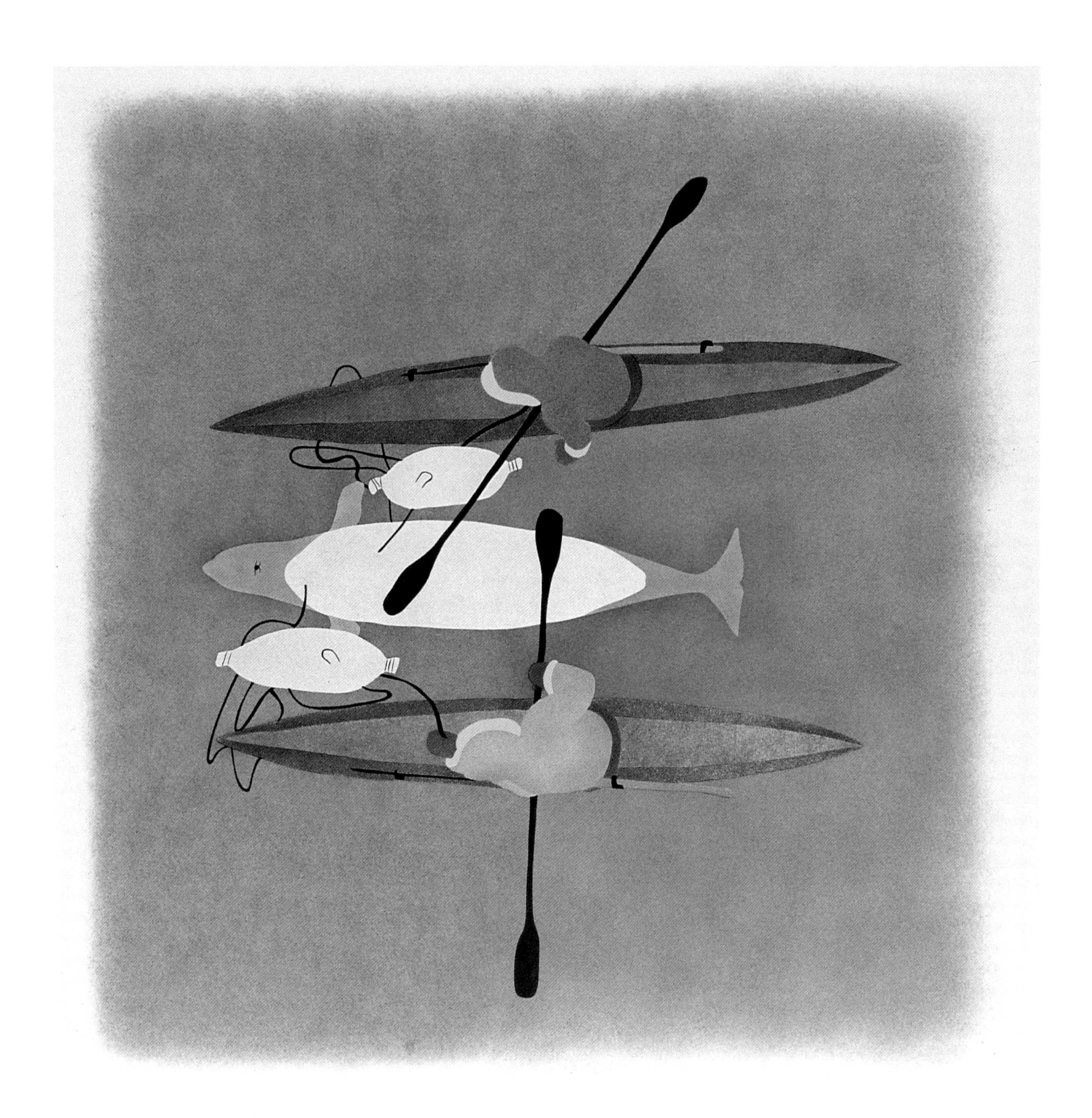

PART II

PEOPLE AND THE ARCTIC WHALES

Chapter 5 EARLY INUIT WHALING

HUNTING THE NARWHAL

The hunter cups his browned, rugged hands around his eyes to shield them from the side glare as he watches the water, ruffled and blue, spread below his rocky vantage point. He sits comfortably on his heels, loose high-top skin boots smooth and tan against his near-white fur leggings. He may be here all day, and tomorrow too. He will be here until he sees a smoky puff on the water's surface for the seconds it exists before it is swallowed by the July sun and cold wind. Then, at one moment on one day, he will see it as he has seen it for the past twenty summers, every summer since his sixth year when he sat on this very spot with his father and learned the hunter's patience, how to watch, how to listen. On the beach below, a dozen flimsy one-man skin boats, kayaks, have been ready for more than two weeks. Lances, harpoons and coiled rope are lashed to the bow of each craft. The rope's end is attached aft to a float made by inflating a seal's skin. The hunters are ready. They are waiting.

Creases deepen on the hunter's broad, sun-browned face. His dark eyes flicker. He is as still as stone. Suddenly he leaps to his feet and cups his hands to his mouth, shouting a signal as he leaps downhill over black granite boulders and dried clumps of hard soil to the beach: "Whales!" And then: "The whales are here!"

Excitement explodes on the beach. Young women with babies frightened by the sudden noise, old women with anxious eyes and a miscellany of children jump and shout and run after the hunters as they stream down the beach carrying their oiled skin boats like surf-

FACING PAGE: *This print, entitled* The Whale Hunt, *by Mary Pudlat of Cape Dorset illustrates the blend of traditional and modern equipment used by Inuit whalers. The centuries-old design of the sealskin floats and skin boats remain, while rifles have been added.* COURTESY OF INDIAN AND NORTHERN AFFAIRS CANADA. REPRODUCED BY PERMISSION OF THE ARTIST.

boards under their arms. In one fluid movement, the hunter slides his boat into the sea, slips into its shallow well, and begins to paddle with the powerful rhythm that will send his tiny craft skimming over the surface fast enough to overtake the whales. If successful, the hunters will harpoon a narwhal and tonight will be a night of celebration with muktuk, the delicious whale skin and blubber, meat to dry for the winter and cache for the dogs, oil from blubber for light and fuel, and perhaps a long tusk for a tent pole or sled runner.

It has been like this for a thousand years in the Arctic community on Inglefield Bay, northwest Greenland. For centuries these northern people have depended on sea mammals for survival. Without them—the whales, seals and walrus—people could never have survived in the Arctic. Not because it is an environment of perpetual ice and snow, because it is not. Not because it is colder than anywhere else on earth, because for periods it is not. It is the persistent cold, the result of short summers, when temperatures average 10°C (50°F) and long, cold winters, when the average annual temperature is –30°C (–22°F). Plants that in other climates provide food, fuel and raw materials for tools simply cannot get water and nutrients from the Arctic's frozen earth.

THIS PAGE: *Two men prepare the highly nutritious and traditional delicacy known as muktuk, which is made from the skin and surface fat of a beluga or narwhal.*
JENNIFER WALTON

FACING PAGE: *Until early in this century, sea mammals such as this dead narwhal provided not only meat and blubber for food but also oil used as fuel for cooking and in stone lamps.*

BOUNTIFUL WHALES

The first Arctic settlers probably arrived from the direction of the Bering Sea, far to the west. Some say they appeared around 1800 BC. Perhaps they already had the technology to whale, or perhaps the whale bones in their middens came from whales washed up dead on the beach, victims of another hunter—the killer whale, perhaps.

Another group, known as the Thule (from the Greek and Roman, "most northerly land in the world") culture, appeared somewhere around AD 900. They moved quickly from west to east, and by the twelfth century they had reached Greenland. They left no written record and few surviving artifacts, since nearly all their household goods originated from animals and therefore decomposed. Thus, it is difficult to piece together the lives of these people who lived so long ago. Yet they were there. They lived, they suffered, they loved, they died, and others took their place. Not only did they survive, they persisted, from Point Barrow in the west to Greenland in the east, an 11 000-km (6835-mile) ribbon of people of a homogenous culture—all because of the Arctic sea mammals, especially the whales.

Where it was possible to hunt the bowhead whale—for example, in the Western Arctic—one bowhead could provide 15 000 kg (33,000 pounds) of good meat and muktuk and 9000 kg (almost 20,000 pounds) of blubber—enough meat for food and enough blubber to burn for light and fuel for a community of five families for an entire season. With such a supply of meat, communities could form permanent settlements close to shore. This meant constructing permanent dwellings, and again the whale provided well. Its huge ribs framed the semi-subterranean houses. Whale skin and sod covered the ribs to form walls and ceiling. Baleen from the whale's mouth was fashioned into baskets, pails, toboggan runners or snares to catch birds and small land mammals. All the community needed was one large whale to make the difference between a winter of food and security and one of hardship and need.

To get the whale, the Inuit made boats by lashing driftwood and animal bones together in a frame that was then covered and sewn as tight as a fine kid glove with sea mammal skin. Sometimes sea lion or seal skin was used, sometimes the skin of a female walrus (skins of male walrus, though more than twice the size, were apt to be scarred and punctured from fierce territorial battles) and sometimes the skin of a beluga. The type of skin, the size of the vessel and how it was used varied according to need. One-man boats, kayaks, were ideal for quick travel, whereas two-man versions were preferred for hunting walrus; one person could paddle and the other could throw the lance or harpoon. Large open boats known as umiaks could transport an entire Inuit family with everything it needed for a season at the summer hunting and fishing grounds, and in the spring, it could carry a crew to hunt the big whale.

FACING PAGE: *Successful whaling meant food in abundance for Inuit families.* PHOTOGRAPH BY GEORGE COMER. REPRODUCED BY PERMISSION OF THE OLD DARTMOUTH HISTORICAL SOCIETY—NEW BEDFORD WHALING MUSEUM.

HUNTING THE HUGE BOWHEAD

Like the middle of an hourglass, Bering Strait is the neck through which water and whales must move during the migration between the Chukchi and Beaufort Seas in the north and the Bering Sea in the south. Between the shoulders of land, the passage is choked with ice most of the year, leaving only narrow open-water leads for the whales to pass through. They swim through the watery corridor, single file, head to tail, tens, hundreds, thousands, passing like ducks in a shooting gallery. It was this combination of climate and geography that conspired to lay the mammoth water beasts at the feet of ancient hunters in this region.

The earliest Bering Strait whale hunters may have used darts poisoned with a plant derivative, a technique borrowed from the Aleuts of the Aleutian Islands to the east. In his small skin boat, the master seaman paddled silently behind the whale, closing the gap between boat and prey until with one mighty sweep of his paddle he propelled his boat beside the whale and threw his poison-tipped lance into the side of the whale. If the hunter threw with accuracy and strength and pierced the body cavity, the whale would be dead in a few days. If the throw fell short and hit the tail flukes, the whale might live for a week or more. Then it was a time to wait and hope—wait for the whale to die before it swam too far, and hope that wind and tide would not carry the precious floating carcass beyond reach.

So great a prize motivated hunters to refine both technique and equipment. An arsenal of sophisticated weaponry evolved to include lances tipped with bone, ivory, slate or stone and complicated toggle-headed harpoons attached to skin floats by long lengths of flexible line. Less and less the death of the whale was left to chance.

In Greenland or in Bering Strait, adult holders of status, wealth and power in the community were the whalers, and every young man wanted to grow up to be one. Each hoped that one day he would have the strong legs, sturdy torso, powerful arms, sharp eyes and exquisite co-ordination to become a master harpooner. Each assumed he possessed the courage to face up to 100 tonnes (98 tons) and 18 m (60 feet) of thrashing fury from a flimsy skin boat.

Whales were hunted in open water in large umiaks manned by a crew of three or four paddlers, a helmsman and a harpooner. On the trail of a whale, the whalers dig their paddles deep into the sloppy sea, and pulling hard again and again, follow the path of the whale as it swims below the surface, anticipating when and where it will rise to breathe. One man stands ready, the heavy harpoon above his right shoulder clenched in thick, cracked hands. As the paddlers manoeuvre the boat beside the rising whale, the harpooner lurches forward as the harpoon leaves his hands and the air leaves his body with the force of his throw. Jerk-

FACING PAGE: The Bowhead Whale, *a print by Peter Aliknak and Susie Malgokak of Holman Island, shows an Inuit hunting polar bears that are scavenging at the site of a partially butchered bowhead whale. Following a successful hunt, the entire community was needed to prepare the tonnes of meat, blubber, bone and baleen.* COURTESY OF INDIAN AND NORTHERN AFFAIRS CANADA. REPRODUCED BY PERMISSION OF THE ARTIST.

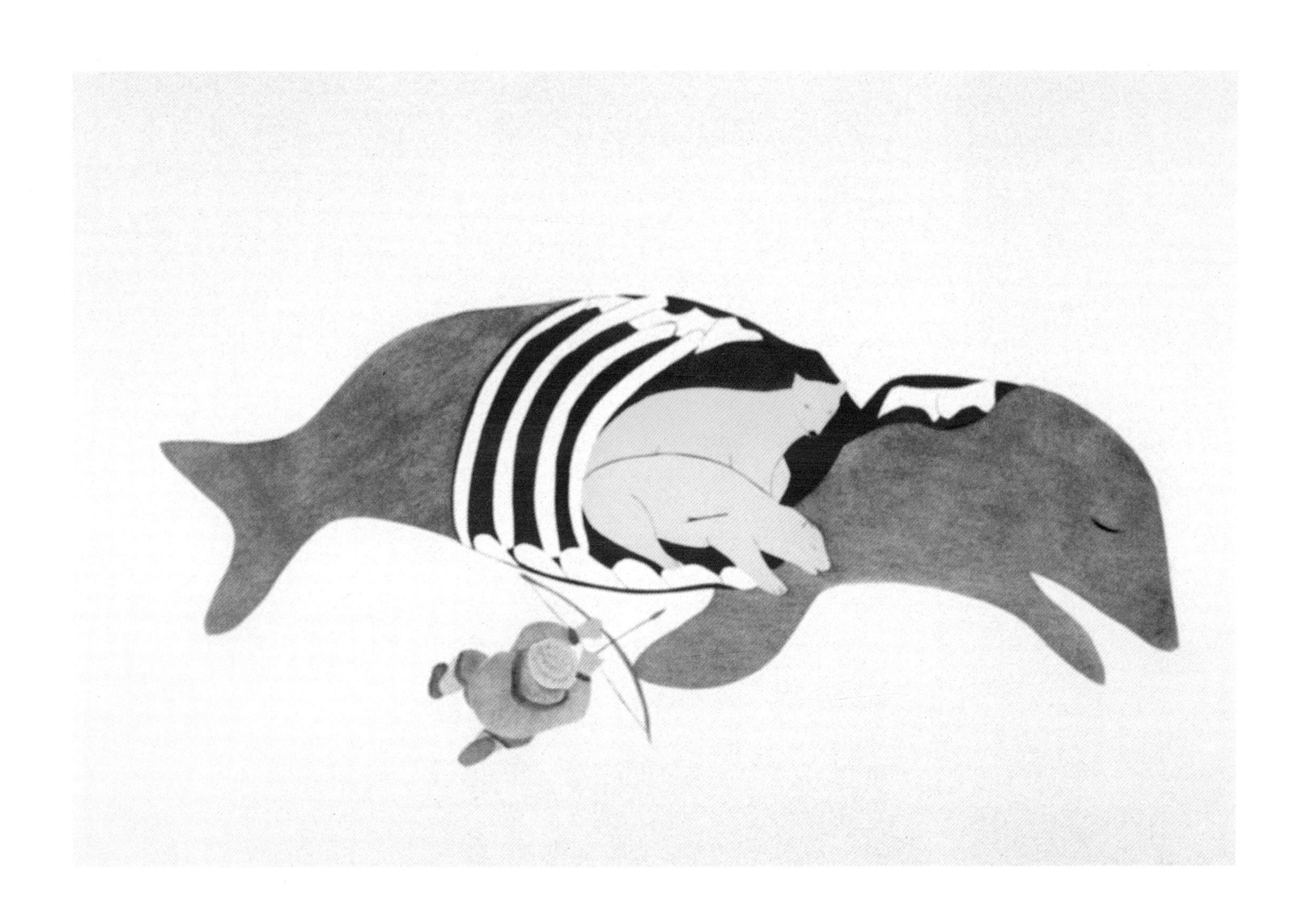

ing rope follows the arching harpoon as it rockets to the wall of glistening black skin. The whale is so close that the hunters can smell its oily muskiness. A perfect hit! The harpoon strikes home, its 25-cm (10-inch) head slicing into the whale below the pectoral flipper. Stung and surprised, the whale dives to escape the pain. Thick, handmade rope streams from the coil in the bottom of the boat, floats of inflated sealskin bounce over the gunwale, and the whalers paddle for their lives before the exposed 8-m (26-foot)-wide tail flukes crash onto the ocean's surface.

The harpooner is a whaler of much experience. He knows where the whale will surface and signals his paddlers to the spot. Other boats see the sign and close in. They wait—tense, silent, tasting the danger, heavy lances raised and ready. A mound of smooth water rises from the sea in the instant before the huge black mass rockets half out of the water in its rage and pain. Lances stab and stab and stab. With the whale's every movement the sealskin float pulls at the harpoon head lodged in its side, gnawing and cutting the whale's huge, tender organs. Again and again, the whale sounds, diving deep, but the relentless drag of the sealskin float twists the harpoon head and chews its insides. Each time the whale surfaces, it's cut with more and more lances. Warm blood and years of life seep into the sea. Red foam boils from the whale's blowhole as it rises for the last time. The whale surrenders and dies, released at last from the hours of agony.

Death to the whale; success to the hunters. The whale boats, tied to each other head to tail, tow the whale to a butchering place.

The whale brought wealth to the community. Wealth begat leisure and the time to create and refine harpoon heads of various configurations and materials, some for whales, some for seals, and gadgets of all kinds—swivels for line, plugs for floats, snow picks, combs and goggles. The bountiful whale meant a lot of food for lots of people, and so the great whaling villages developed at Point Hope, Point Barrow and other places, where successful whaling depended on large, stable villages for manpower and provided for large groups of people to live in concentrated locations. More people meant more whaling.

ABOVE: *Land-based activities such as fishing, travel and child care are represented on the left side of this Inuit tapestry. The Arctic whales are depicted under water to the right.* JENNIFER WALTON

FACING PAGE: *The ivory gull is a year-round Arctic resident along with the Inuit and the Arctic whales.*

According to archaeological evidence, the great whaling villages finally collapsed. Perhaps it was climate change and an alteration in the animal's migration patterns, as some have suggested, or perhaps decades of ever-increasing hunting pressure on the whales finally overwhelmed their capacity for renewal. Had the numbers of whales killed slowly increased until those killed outnumbered the young being born? Perhaps as in so many other places the hunters believed the animals would keep coming back forever. Some believe that overhunting is a by-product of "modern" weapons. Not so. Weapons don't overhunt, people do. It just takes a little longer with "traditional" weapons.

It would be incorrect to suggest that northern hunters hunted only whales, because they did not, though a detailed picture of how the northern hunters made their living is still incomplete. According to the season and location, they fished for fish and shellfish and hunted caribou, birds and marine mammals, including the bowhead, narwhal and beluga whales. They adapted techniques according to the animal and its location. For example, when hunting belugas, they sometimes whaled from the edge of the ice, they sometimes drove the belugas into shallow river flats, and they sometimes hunted the whales from a boat.

We do know that the Inuit hunters were highly skilled, appropriately clothed and housed for the cold climate and extraordinarily mobile with their sled and dogs for travel over the ice in winter and with their boats in summer. For centuries they alone exploited the Arctic's rich marine resources. Perhaps it was inevitable that one day another hunter in a bigger boat would come from far across the sea.

FACING PAGE: *Highly mobile with their skin boats in the summer and dog sleds in the winter, northern people not only survived but thrived for hundreds of years in what appears to most outsiders to be a hostile and forbidding land.*

The narwhal was the only Arctic whale not to suffer irreparable damage from overhunting in the early days of commercial whaling.

"EOTHEN."
"ISABELLA."
"A. LAWRENCE."
"A. BRADFORD."

Chapter 6 COMMERCIAL WHALING

Arctic subsistence hunters were not the only whalers. In Europe, "Northmen" were whaling in the ninth century, and the Basques from the Bay of Biscay fished for whales well before the invention of the mariner's compass. By the thirteenth century, Basques and whalers from coastal towns all over Europe were whaling hard on the Atlantic right whale, pressing north as whales vanished as a result of overfishing in the south.

BOWHEAD WHALING

With each season, whalers sailed farther and farther north in their relentless search for unexploited whale stocks. By the fourteenth century, they were well past 75° north latitude in the icy Barents Sea, where, near Spitsbergen, they discovered a new kind of whale, which was similar to the greatly prized right whale. It was as big as the right whale, if not bigger, slow and fat, its baleen was longer and finer than that of the right whale, and it floated when dead, an important consideration for the early whalers, who lacked the manpower and machinery to winch tonnes of dead whale from the sea bottom. Each nationality that hunted this whale called it by a different name, but eventually it became known as the great polar whale or the Greenland whale. Today it is known as the bowhead whale.

The discovery of the bowhead whale drew whale ships from all over coastal Europe: the French, Dutch, Danes, Germans and British all came for season after season of lucrative

FACING PAGE: *To save travelling time to and from the rich Arctic whaling grounds, some whalemen overwintered in the north. Inuit hunters were engaged to provide fresh meat for the ice-bound crew.* REPRODUCED BY PERMISSION OF THE OLD DARTMOUTH HISTORICAL SOCIETY—NEW BEDFORD WHALING MUSEUM.

whaling. By the late 1600s, the Dutch alone had three hundred ships and eighteen thousand men working around the clock in the twenty-four-hour daylight of the Arctic summer, cooking whale blubber to release the oil, which was then casked, brought back to Europe and sold as lamp fuel and lubricant and for use in the manufacture of soft soaps, varnish, leather and woollen goods. Thousands upon thousands of bowhead whales were killed for their oil; 100 barrels could be obtained from an average-sized whale and up to 300 barrels from a really big one. In addition, each whale yielded some 680 kg (1500 pounds) of baleen, which the whalers called whalebone. It was light, flexible and unmatched by any other stiffening material for umbrellas, corset stays, petticoat hoops, hats and furniture springs. Vessel owners, investors, captains and crews made fortunes. Then, almost overnight, there were not enough whales to cover the cost of the trip north.

By the early eighteenth century, the whalers moved from Spitsbergen to Iceland, to the east coast of Greenland, around the southern tip of Greenland and, by 1719, into Davis Strait between Greenland and Baffin Island. Within a few seasons, 350 ships were on the grounds, all powered by sail and oar, all whaling hard between the ice breakup in the spring and its return in the fall. In the first ten years of the Davis Strait fishery, the Dutch alone sent 747 ships to kill whales. Records show 5886 ships taking 32,907 whales during the forty-six years ending in 1721. Although the Dutch were major players, they were not alone. The slaughter was enormous. Whalers fanned out from Davis Strait into Baffin Bay, Lancaster Sound, Foxe Basin and Hudson Bay. They set up semipermanent shore stations with sheds, cookers and places to cut up and render the blubber. When the extracted oil cooled, it was stored in barrels and finally rolled into the sea and towed to the ship for the return trip to Europe.

In some areas the relationship between whalers and Inuit was so close that Inuit lived on board the whale ships for all or part of the season. Here a group of adults and children is photographed in the ship's cabin. PHOTOGRAPH BY GEORGE COMER. REPRODUCED BY PERMISSION OF THE OLD DARTMOUTH HISTORICAL SOCIETY—NEW BEDFORD WHALING MUSEUM.

The Inuit worked in the whaling operations as oarsmen and harpooners. Sometimes their wives and families lived on board the whale ship for the season. Local people carried on active commerce with the whale ships, supplying them with fresh meat and skins from bear, wolf, seal and walrus, as well as ivory from walrus and narwhal tusks. In trade, they took metal merchandise such as cooking pots and tools.

It was a lively and expansive time all around, but it was all over in less than fifty years. By the mid-18th century, the bowhead whale fishery in the North Pacific had collapsed.

"Fishery" is the term used for hunting and killing whales, yet the only feature whales and fish share is that they swim in the ocean. Whales, late to mature and slow to reproduce, give birth to a single live young only every two or three years. With such a low replacement rate, it is no surprise that a "whale fishery" will crash with predictable certitude once the whalers start "fishing."

For centuries, whaling had been a feast-or-famine proposition; when one area dried up, the race was on to find a new location with unexploited whale stocks. In the mid-eighteenth century, bowheads were gone from Davis Strait, but the "fishery" was not over yet.

In search of their prey, the whalers moved out of the North Atlantic, into the South Atlantic, around the Horn, up the coast of Chile and into the Pacific by 1791. Seven hundred vessels joined in the trade. Yankee whalers from Nantucket, New Bedford and other east coast ports, as well as ships now sailing out of San Francisco on the west coast, pushed farther north every year, chasing the gray whale, sperm whale and North Pacific right whale. One east coast whaler working in the North Pacific, Captain Thomas Roys of Sag Harbor, Long Island, was a bold and adventurous mariner. He took his whale ship, the *Superior*, through Bering Strait into the Chukchi Sea to become the first recorded whaler to sail into the Arctic. There Roys found the Pacific bowheads. The year was 1848.

In a time when all communications were transmitted by word of mouth or by letter, transoceanic news travelled no faster than a ship could sail. It is therefore truly remarkable how quickly news of Roys's bowhead discovery moved. Following his trip to the Chukchi Sea, Roys sailed into Hawaii, and within three years of his arrival there, two hundred vessels were in Bering Strait harpooning bowhead whales. With each season, the ships probed deeper into the Arctic through the strait, east to Beachey Point and, by 1887, to the MacKenzie Delta. To save travel time, some whale ships overwintered at Herschel Island, deliberately becoming frozen in the ice, in order to be on the whale grounds even before the whales arrived in the spring.

Tales of North Pacific whaling are full of suffering and hardship. Men left their families for up to five years. Some died, victims of disease, drowned when wounded whales crushed small whale boats like so many matchsticks or burned to death when shipboard tryworks spilled boiling oil or sent the entire ship up in flames. One hundred and fifty ships were lost during the seventy years of Arctic bowhead whaling, sunk when their bellies' ripped open on ragged shoals hidden by fog, or when their hulls were crushed like paper in fields of moving ice. Hundreds of lives were lost.

Something else was lost too—bowhead whales, tens of thousands of the great, black magnificent beasts, including countless female bowheads and thousands of their calves. As the whales slowly died, thousands upon thousands of litres of whale blood flowed into the sea. Hundreds of thousands of kilograms of blubber-stripped carcasses were left to rot. Still the hunt went on.

A 1905 photograph of men holding bundles of "whalebone," which is not bone at all but the baleen plates that hang from the mouth of a baleen whale. Whalebone is strong, light and flexible, an excellent material for furniture springs, corset stays and umbrella ribs. REPRODUCED BY PERMISSION OF THE OLD DARTMOUTH HISTORICAL SOCIETY—NEW BEDFORD WHALING MUSEUM.

Smaller, tougher, steam-powered vessels were in use by the 1880s. They followed the whales into the ice, where they had fled to escape the sail-powered whalers. Soon the bomb lance and shoulder gun, an invention of Norwegian Svend Foyn in 1864, came into common use and introduced a new and gruesome chapter in whale history known as the modern whale fishery. Now other species of whales previously beyond the whaler's reach because of their size and speed fell within target range.

Although they were hunted into the turn of the century, bowhead whales do not play a major role in modern whaling. Petroleum, discovered in 1859, substituted for whale oil, and later, spring steel substituted for baleen. Even if these alternatives had not appeared, the sixty to seventy years of Pacific bowhead whaling had brought the bowhead whale to the brink of extirpation. It also brought dramatic changes to the Western Arctic Eskimo, later known as the Inuit. During its most active period, fifteen shore stations operated from Point Barrow to Cape Thompson. These stations needed manpower, and competition for Eskimo crews was fierce. In return for two months of service, whalers offered local workers a year's supply of flour, rifles, ammunition and manufactured goods. Although 1897 was the last good bowhead season in the Western Arctic, bowhead whaling continued for another dozen years.

THE END OF COMMERCIAL BOWHEAD WHALING

We will never know how many bowhead whales there were before the onset of commercial whaling, but we do know that there were plenty. Scientist provide a best estimate based on old whaling records, ships' logs, diaries and naturalists' observations. Sometimes old records of whaling company accounts are used to work backwards and calculate how many whales it would take to produce the recorded barrels of oil and pounds of baleen. Records show 28,000 bowheads killed in the Eastern Arctic between 1791 and 1911. Not included in this number are wounded whales that died and were not recovered, orphaned calves who died after their mothers were killed and whales represented by baleen and oil lost in accidents, shipwrecks or fires. Scientists estimate that there were 11,000 bowheads in the original Davis Strait stock. Today there are about 250. Hudson Bay is estimated to have had somewhere between 450 and 650 bowheads. Today there are a few tens. In Spitsbergen, there were thousands then, today a few tens.

The Bering Sea stock, the last group to be whaled, is estimated to have numbered at least 18,000 to 20,000. How many remained when commercial whaling ended no one knows. Even though there has been no commercial whaling on this stock for some seventy years, in 1979 scientists estimated there were only 2300 whales. Yet by 1992 some scientists were saying that the stock numbered more than 7000. Since such a dramatic increase in so short a time contradicts the bowhead's rate of recovery elsewhere, it is a mystery where all these whales came from. Are they real, or are they paper whales—that is, numbers on paper, created by multiplying the actual number of whales seen by some other number to account for whales that might be there but not seen, swimming under water or hidden by waves, or calves blocked by the bulk of their mothers. It may well be that the threefold increase in numbers is real, the result of better research and more sophisticated counting methods.

Commercial whalers hunted belugas throughout their range for meat, oil and hides.

BELUGA HUNTING

Commercial whalers hunted other Arctic whales as well. Throughout their range, belugas were hunted, trapped and netted in most areas where they occurred. They were taken sporadically around Spitsbergen in the early days and then as a regular fishery between 1867 and 1878. That population seems not to have recovered. The former Soviet Union continues to hunt belugas; it took at least five thousand between 1969 and 1980. In the Gulf of Saint Lawrence a population of 5000 belugas in 1885 was reduced to a few hundred in ninety-three years of hunting, which included a Government of Quebec–sponsored bounty hunt in the 1930s with $15 paid for each whale killed to "protect" the cod fishery. Two thousand bounties were paid. In 1983 the St. Lawrence beluga was declared endangered.

Although belugas in the Saint Lawrence were granted legal protection from hunting in 1978, more than twenty years of protection has not increased their numbers. Dams, pollution, disturbance and interference in the food supply have replaced hunters as hazards to survival.

Commercial hunts for oil—and later, hides—began in 1689, when twenty-eight casks of beluga oil left the Hudson's Bay Company's post at Churchill, Hudson Bay, for England. Every trading post situated near a concentration of belugas, which occurred at just about every Arctic river estuary, whaled for belugas by driving the animals into shallow water, where they could be netted, shot or harpooned. Scottish and American commercial whaling vessels were hard at it too, taking more than twenty thousand whales from Davis Strait and Lancaster Sound between 1868 and 1911. The hides were said to make superior boot laces and coach covers.

In northern Manitoba, Churchill on Hudson Bay was back in the beluga business in 1949, turning minced beluga into mink food for prairie fur ranches. Local Inuit, Cree and Métis hunted the whales and were paid by the foot, rather than the pound, for what they brought in. Then, for a short time starting in 1961, a native-run plant produced canned and frozen muktuk. The discovery of high mercury levels in the meat forced the operation to close in 1970.

THE NARWHAL TUSK TRADE

Narwhal tusks entered trade a very, very long time ago. The northern native hunters hunted the narwhal and used the tusks instead of wood, which was not available, for sled runners, tent poles and lance shafts. Thus, they would have had tusks in their possession when early European traders, probably looking for furs, entered the circumpolar world. That the tusk was the tooth of an Arctic whale either was unknown or was a jealously guarded trade secret in order to preserve the product's value, since the tusk was marketed not as a whale tooth but as the priceless horn of the unicorn!

A medieval favourite, the unicorn embodied all that one could hope for on this earth—power and virtue, earthly love, heavenly love, boundless freedom, might, right and nobility. It was hugely valued not only as a decorative ornament but also for its medicinal powers. When powdered and ingested, it gave protection against stomach trouble and epilepsy. Its grander restorative powers endowed the ingester with the dream of every man—health, youth and virility.

A drinking cup or powdered potion of unicorn horn was believed to be the ultimate protection against poison. This was a good market strategy, given that before the advent of firearms and modern forensic medicine, poison was the method of choice to dispatch enemies, rivals and unwanted spouses, parents and siblings. If the poisoner knew that the potential victim had some unicorn horn, then there was no sense in using poison. Thus, in a roundabout way the unicorn horn did provide protection.

Eventually the truth was revealed. When European whalemen, in their relentless search for new and unexploited whaling grounds, followed the bowhead whales to Davis Strait, they found not only bowheads but the "unicorn of the sea," the narwhal. Tusks were still taken in trade from the Inuit, but now they were marketed as high-quality, fine-grained ivory, not unicorn's horns. Events have come full circle, and today the narwhal's tusk is marketed once again as a curio. It is valued not because it comes from a mythical beast but because it comes from a unique and living whale.

Narwhals were hunted commercially on and off from the seventeenth to the twentieth century. Even though the numbers taken were high—10,970 in the Baffin Bay/Davis Strait areas between 1915 and 1924—the lack of a sustained commercial hunt has meant that this species did not suffer the same fate from commercial whale hunters as the bowhead and some beluga populations.

With their predictable return each summer to shallow river estuaries, belugas were easily killed in nets, weirs and drive fisheries.

Chapter 7 THE ARCTIC WHALES TODAY

It would be wonderful to say that with the end of commercial whaling the Arctic whales were finally left in peace to live their extraordinary lives in their unique but demanding environment. Sadly, this cannot be said, because not only does hunting continue but twentieth-century technology is providing ever-increasing opportunities for industrial development in the frozen North such as mining, hydroelectric projects, shore development, shipping and Arctic fishing operations. All of these and even northern whale watching can have a negative impact on the whales and their habitat. In addition to these pressures is the very real challenge of rehabilitating many stocks already depleted by past commercial exploitation.

There is a generally accepted notion that whales are protected by international regulation. So, if asked, most urban North Americans would say there is no whaling today in Canada or the United States. They would be wrong. Americans hunt bowheads and belugas; Canadians hunt both, plus the narwhal. Greenlanders and people of the Chukotka hunt Arctic whales as well. Moreover "traditional" hunting methods are being replaced by the use of outboard motors, high-powered rifles, explosive harpoons and cellular phones to coordinate the hunts. There is more hunting, more waste and in some cases declining stocks. Whale products are entering commerce in a limited way.

FACING PAGE: *The face of the Arctic is changing as the human presence there increases.*

QUOTAS

In 1982 the International Whaling Commission (IWC) voted for a complete moratorium on commercial whaling of large whales until a new management regime was developed. Aboriginal subsistence whaling, however, would continue within quotas set by the IWC. But only nations that are members of the IWC are bound by IWC regulations, and countries can join or not as they choose. Belugas and narwhals are classified as "small cetaceans" and as such are not subject to IWC regulation but are the responsibility of the nations in whose waters they occur. This is how whaling of Arctic species continues even though "commercial" whaling is still under moratorium.

In the United States, the bowhead whale is designated an endangered species. Aboriginal hunting is allowed, however. Bowheads are hunted by less than a dozen whaling villages in Alaska under a three-year quota, which, for the years 1992 to 1994, permitted 141 strikes and allowed up to 41 whales to be landed each year. Since 1978, there has been a gradual increase in the number of bowhead whales allowed to be struck and/or landed, so the quota is now almost three times what it was. In Canada, the Western Arctic Inuvialuit community of Aklavik was granted a licence in 1991 by the Canadian minister of fisheries to kill a bowhead whale, the first in decades. The Aklavik whale came from the same Chukchi/Beaufort Sea stock as do the bowhead whales hunted by the Alaskans, but it was not sanctioned by an IWC quota, since Canada has not been a member since 1982. In Canada, the Inuvialuit Land Claims Settlement of 1984 in the Western Arctic and the Nunavut Land Claims Agreement of 1993 in the Eastern Arctic grant the people of these areas the right to hunt all wildlife in the settlement areas, subject only to restrictions related to conservation and human health and safety.

Following the Canadian Inuit bowhead hunt for one whale, the Americans brought considerable diplomatic pressure to bear on Canada, citing the impact of the Canadian hunt on the "conservation" of the bowhead whale. The protestations ring as hollow as a skin drum, given the endangered status of the whale, the Alaskans' fifteen-year escalation of the hunt and the generous quota allocation from the IWC.

Scientists say that the beluga stocks shared by Greenland and eastern Canada have been overhunted and are now half of what they were in 1981, and that if hunting is not severely curtailed, the whales will be gone by the year 2004. Examination of dead whales from large groups driven into shallow water and killed reveal that 35 per cent were pregnant females. This is a huge proportion, considering belugas bear young once every third year. Alaskan natives take a couple of hundred belugas each year from the Western Arctic with no nega-

A rare double-tusked narwhal killed in the subsistence whale hunt. Northern communities are being challenged to find new ways to manage subsistence hunting to avoid overexploitation.

tive impact on the population there. The total number of belugas taken throughout the Canadian Arctic ranged from 1300 to 700 between 1974 and 1986. Today populations in Cumberland Sound and Ungava Bay are endangered, the eastern Hudson Bay population is classified as threatened, and the Western Arctic population is healthy.

Scientists believe that narwhal stocks shared by Greenland and eastern Canada are in serious decline. A 1979 study of a narwhal hunt estimated that at least half the animals that were shot sank and were lost. Others were wounded and probably died later. Many of those killed were pregnant females. Suckling young died when their mothers were killed. Some scientists estimate that although 552 narwhal kills were reported by the hunters, the death toll may have been closer to 2000.

A quota of 527 narwhals per year has been established for the Inuit of the Canadian Arctic, with an average estimated kill of 450 to 500 per year. West Greenland Inuit averaged a take of 800 per year during the 1980s. Greenland has no quotas.

Soapstone carving of a male narwhal from Pangnirtung on Baffin Island by Philiposee Kuniliusee. COURTESY OF INDIAN AND NORTHERN AFFAIRS CANADA. REPRODUCED BY PERMISSION OF THE ARTIST.

WHALERS VERSUS ANTIWHALERS

When quotas are placed on bowheads in Alaska and Canada, narwhals in Greenland, and belugas wherever they are hunted, these quotas are typically related not to conservation but to food and to social and ceremonial needs. Aboriginal people argue their right to hunt whales based on history, a special relationship with wildlife and cultural necessity. Those opposed to whaling have similarly emotional arguments, which, again, centre on human needs. Some antiwhalers accord whales special status in the animal world and object to their being killed, even for food, since there are readily available substitutes for everything a whale provides. Some elevate whales (no distinction between species) to the level of superanimal, almost superhuman, as loving, social, caring, gentle and intelligent beings. To these people, their relationship with whales is spiritual too. They revere the whale as an ideal of humankind's relationship with living things. Others object to whaling on humane grounds. In truth the arguments of both the hunters and the preservationists have little to do with whales and have everything to do with people's feelings, beliefs and culture.

For the whale, dead is dead. The whale is no less dead if it is killed for cultural or traditional reasons than if it is boiled for lamp oil.

The whalers and the antiwhalers lock horns with each other as they argue on moral, cultural, ethical and historical grounds. Each side hires expensive legal council to thwart the other. Conservationists enter the fray, with the position that if whales are going to be hunted, then limits must be set to avoid overexploitation. Ironically, it may be better for the whales that this conflict not be resolved. The debate forces management authorities to establish defensible limits on the numbers of animals killed. This means that research must be carried out to determine the whales' numbers, biology, natural history and travel patterns. This need for information provides the motivation for both sides to spend the time and money needed to study the Arctic whales.

While gaps in information are being filled, there is a strong argument for conservative management. It's time to err on the side of the whales, or there simply won't be any. There's no use having a right to something that doesn't exist.

HUMAN DISTURBANCE

Early on June 23, 1983, a thousand belugas and more than a hundred narwhals cruise leisurely along the front edge of an ice sheet blocking the entrance to Lancaster Sound. Ice breakup is late this year, and the whales swim back and forth along the ice edge from Bylot Island on northern Baffin north across the sound to Devon Island, searching for a crack in the impenetrable ice as they wait for it to falter, crumble and finally allow belugas passage into the high Arctic archipelago and narwhals to move into Admiralty Inlet, where they will spend the brief Arctic summer. As long as the ice holds, the whales are held back, milling, like so many sprinters who wait for the starting gun. Some are impatient. They've waited for a week already.

Under water, the sea is saturated with the whistles, squawks, barks, blaring, trills, click-clatter and pulsed calls of whales and bearded seals. At 8:00 A.M. there is a noticeable difference in the character of the calls.

More than an hour passes. Suddenly the invisible starting gun goes off, though the ice is as solid as ever. White backs roll high in the water as belugas stampede along the ice edge, tail flukes pumping in frantic efforts to escape. The narwhals don't move, paralyzed by fear. Small groups huddle together silently, so close they touch. They sink silently downward. It's as if they want to be very quiet and very still, in hopes they won't be noticed.

Two hours pass. All is quiet at the ice edge. The whales are gone, and snow is falling now in a gossamer curtain. Half an hour later a huge grey form begins to take shape in the cloud of falling snow, gaining definition as it advances over the water through the heavy snow shower. It is the MV *Arctic*, an ice-breaking ore carrier on its way to the lead-zinc mines at Nanisivik on Admiralty Inlet. It seems to take forever for the ship to emerge as a complete shape in the falling snow. At 209 m (686 feet), it could span the length of two football fields. Its great belly hangs 11 m (36 feet) below the surface, equivalent to a four-storey building. Its size is described in shipping terms as 20 000 dead weight tonnes, and it is pushed through the water by a gigantic single 14,770-horsepower diesel engine.

Underwater sound-sensing instruments picked up the ship when it was 132 km (82 miles) away from the ice edge. The whales may have heard it long before that, but they didn't show concern until it was 84 km (52 miles) away. Imagine hearing something so far away that to drive the distance would take over an hour and forty minutes at 60 km/h (40 miles per hour). When the ship was still 40 to 55 km (25 to 34 miles) away, the belugas panicked and fled. An hour and a half later they were 49 km (30 miles) north of the ice edge and still moving.

FACING PAGE: *It has not yet been determined to what degree the underwater sounds of ships engines, propellers and ice-breaking activities interfere with the whales' ability to communicate, navigate and find food in Arctic seas.*

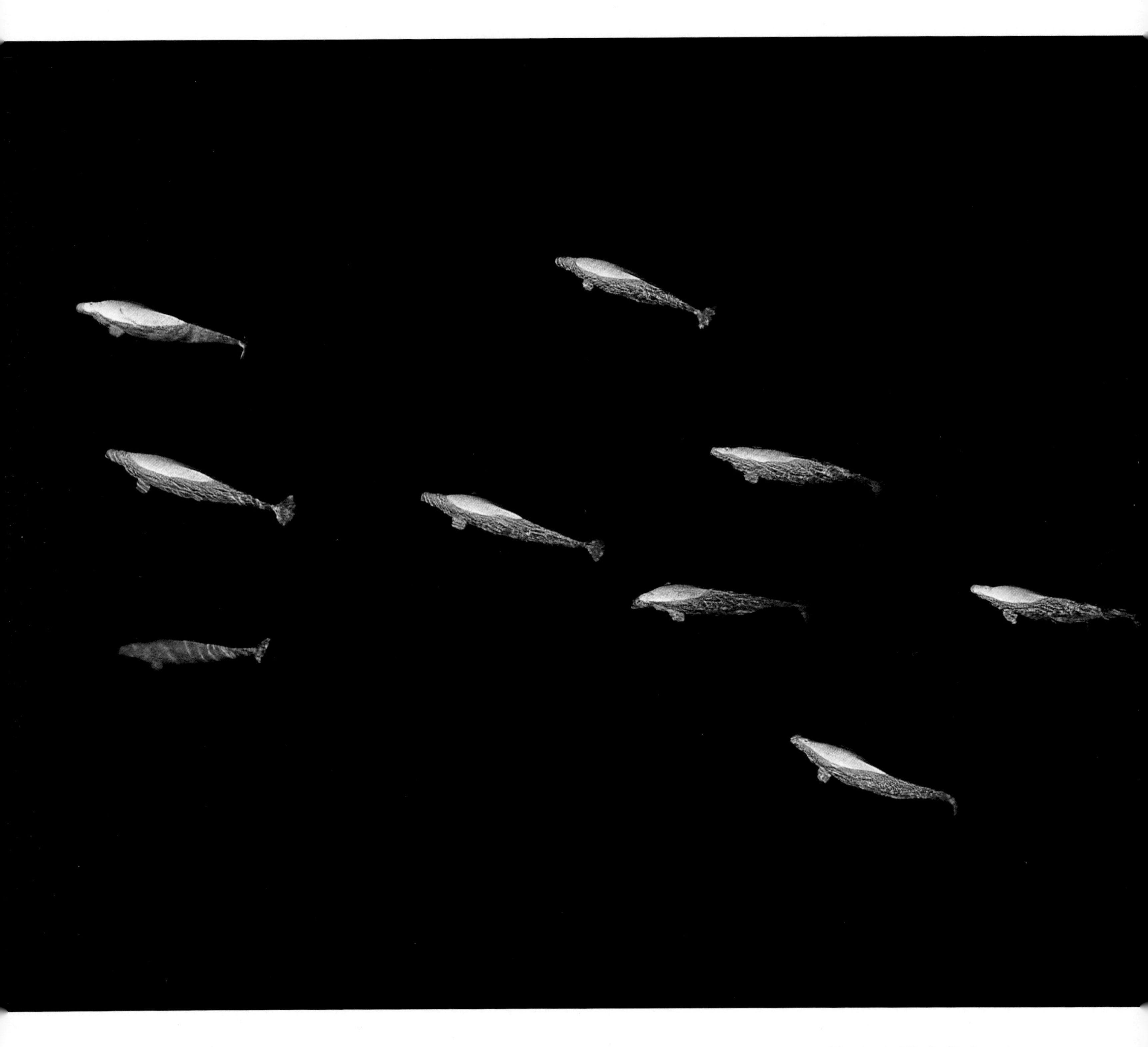

Ice-breaking and engine noise from the *Arctic* and the Canadian Coast Guard icebreaker the *John A. MacDonald*, arriving two days after the *Arctic*, displaced whales up to 80 km (50 miles) away from where the ships entered the ice. Within a day or less of the ship's passing, many whales returned. Even though the ships were 10 km (6 miles) into the ice when the engines started up again, after a day of silence, the whales fled once more.

Scientists set up camp on the shores of Lancaster Sound in late June of 1982, 1983 and 1984 to observe the effect of shipping on whales that had not previously been exposed to heavy vessel traffic. The study was undertaken when native narwhal hunters expressed concern that a proposal to increase the commercial shipping season to and from the mine at Nanasivik by breaking through the ice of Admiralty Inlet would frighten the whales. They were right. Observations over the three periods clearly demonstrate that Lancaster Sound whales were dramatically affected by the noise and/or presence of ice-breaking ships. Some theoretical calculations on the hearing abilities of belugas tumbled into the sea as a result. One calculation predicted that belugas would hear the ship at 5 km (3 miles). This was subsequently revised to 20 km (12 miles). The reality is that the whales heard the vessel when it was at least 84 km (52 miles) away, meaning that the range of disturbance from ships is far, far greater than anyone had imagined.

It is not as simple as "boats bother whales." Detours and displacement cost the whales time and energy. Imagine that each day you walk a kilometre to your place of work at a market garden. One day without warning you are forced to detour 3 km (2 miles) out of your way to and from your destination. Your trip to work now takes you three times as long, but you still have to do your job, so you have to make up the extra hours spent travelling out of your own time. Because you are walking, you use up three times as much energy getting there, meaning that you are hungry and need to eat more but you have less time to do it in. If you are a mother with a dependent child, the burden increases because you can only move as fast as your toddler can walk, and the youngster is expending energy in extra travel that would otherwise be used for growth and development. This is essentially what happens when wild animals are displaced. The time they need to hunt for food is reduced at the same time their energy needs increase.

For ships, Lancaster Sound is the eastern entrance to the famous Northwest Passage, the proposed route for year-round shipping of mine products, the route used to transport hydrocarbons from the Western Arctic to markets on the eastern seaboard. It is also the entrance to and the route for people and machinery necessary for industrial development in the high Arctic. For narwhals, belugas, bowhead whales, bearded seals and walrus, Lancaster Sound is the major migration corridor that leads to summering areas and abundant food in the high Arctic archipelago.

Shipping is not only a concern in the Eastern Arctic but also the subject of considerable study in the Western Arctic. Add to vessel engine noise the sound from helicopters, airplanes, drilling, dredging and seismic testing and there is a real concern about how the increase in artificial background noise interferes with the whale's ability to hear the sounds it needs to navigate, communicate and find food.

Since the early 1970s, offshore industrial development has proceeded rapidly in the Western Arctic's Beaufort Sea, principally as a result of oil and gas exploration, which often occurs at the same water depths as offshore areas frequented by migrating bowhead whales. By 1983, bowheads travelling through the area heard five drill ships, five seismic ships, two drilling caissons, four industrial icebreakers, six large dredges, eight supply ships, ten twin-engine helicopters and miscellaneous tugs, barges and supply vessels. Some whales did not seem to be bothered and others avoided the area, but for the most part the effects are not known.

OTHER EFFECTS OF INDUSTRIAL DEVELOPMENT

Coastal shipping, fishing and hunting all create hazards for the whales in the form of noise, collisions, net entanglement and harassment. Toxic pollutants work their way into the Arctic food chain through atmospheric transfer, ocean current and north-flowing rivers. St. Lawrence River belugas are living with the results of past heavy use of agricultural pesticides such as DDT as these poisons flow into the St. Lawrence River from hundreds of kilometres of adjoining countryside. DDT and Mirex levels are up to 100 times greater in St. Lawrence male belugas than in their Arctic cousins. Some toxicologists believe that the levels are high enough to cause partial or complete reproductive failure. This may be part of the reason why in ten years the St. Lawrence belugas show no increase in their numbers.

The belugas' persistent and predictable use of estuaries exposes them to increased risk from overhunting, harassment, disturbance by whale watchers and hydroelectric development, such as the 80 per cent diversion of Western Hudson Bay's Churchill River into the Nelson River.

Liquid natural gas (LNG) reserves in the high Arctic are estimated to be 3 trillion cubic metres (106 trillion cubic feet). It is unlikely that a resource of such magnitude will stay in the ground forever. One proposed LNG tanker was designed with 150,000-shaft horsepower, enough to break through ice that is 3 m (10 feet) thick at a steady 6 knots. Such a vessel would be costly to build and operate, so naturally it would be designed to operate year-round or very nearly so. It's difficult to imagine that this mechanical beast would not have some effect on the whales, not to mention the problems that could arise from creating artificial leads in the ice.

What developments and intrusions into the Arctic there will be in the future can only be guessed at, but we can be sure that where there is profit, there will be development. We must keep in mind, however, that industrial activity is not necessarily a bad thing. It stimulates biological research, albeit typically not at its own initiative, and produces the kind of research money that governments and universities never seem to have. Without the threat of industrial disturbance, there would never be the motivation or money to study the Arctic whales for their own sake. The great irony is that industrial development, which could result in the greatest damage to the Arctic whales, will likely be the source of pressure to understand the behaviour and movements of these unique beasts, their vital parameters and their critical habitats. Armed with such knowledge, industry can, if it chooses, develop around the needs of wildlife.

FACING PAGE: *In addition to hunters, Arctic whales must now contend with developing fisheries, increased shipping and pollutants.*

OUTSIDERS ARE WATCHING

Countries bordering on the Arctic—Canada, the United States, the former Soviet Union, Greenland and the Scandinavian countries—have quietly gone about their business, good and bad, in the Arctic, and for a long time the temperate world neither knew nor cared. But things are changing. The polar regions, both north and south, are undergoing scrutiny from the international environmental community.

Whether appropriate or not, conservationists view unique biological areas such as the Arctic as a kind of global common property, too precious to be left to the management of sovereign states or local groups. Some obvious examples of this kind of thinking are the Amazon rain forest, the Indian Ocean and the Pacific Northwest old-growth rain forests. Some environmental groups are referring to the Arctic as the "New Serengheti" in reference to the area's unique and threatened wildlife, and there are already calls from such bodies as the International Union for the Conservation of Nature (IUCN) to have Isabella Bay in the Canadian Arctic declared a world heritage site.

As you have seen, the Arctic whales are migratory animals, and like all migratory species, all their habitats, including their travel corridors to and from seasonal areas, are equally vital to their survival. The organisms that the whales feed on in their various habitats are crucial as well. Therefore, to be meaningful, habitat protection must extend to the animals' entire range. This is a very tall order.

WHO WILL SPEAK FOR THE WHALES?

There are huge gaps in our knowledge about Arctic whales—how long they live, what they do in the depths of winter, how they choose the summering areas that they do and so on. Yet in the past, without our knowing anything about them, the Arctic whales lived and thrived in their Arctic home for millions of years. They could continue to do so but for the fact that we have entered their world with the power to change it. We cannot know how our activities affect the whales until we know more about the whales themselves, and they are extraordinarily difficult animals to study. Their range is huge. They live in an environment that is cold and dark for much of the year. Even during the short Arctic summer it is both expensive and awkward to get near the whales, since most travel is by chartered aircraft and is always subject to delay because of poor weather conditions. In the field, cold is hard on both researchers and equipment. Something as simple as counting the whales is complicated by weather and the fact that the whales are under water and out of sight 90 per cent of the time.

Learning more about the whales is difficult, but not impossible. Hunters can play an important role in research by making available body measurements, stomach contents, reproductive organs and other tissues for researchers. Samples and measurements can provide researchers with valuable information about diet, reproductive biology, growth rates and pollutants. Satellite radio tags have been attached to a few whales and are providing amazing new information about travel, swimming speeds and diving depth. The satellite radio tag is the size of a loaf of bread and consists of an instrument pack and antenna attached to the whale's back. When the whale is at the surface, the tag sends a radio signal which is then picked up via satellite, relayed to a computer station in Europe and then re-relayed to researchers in Canada. The attachment mechanism is designed to deteriorate and release the pack after a short time. Real information from these instruments is replacing old assumptions such as the long-held belief that belugas were slow, shallow-water, near-shore travellers. The actual behaviour of tagged belugas gives quite a different picture. In another experiment, a narwhal fitted with a similar device in the Eastern Canadian Arctic was followed via satellite all the way to Greenland. This is more than just fascinating information. It has serious application to determining safe hunting levels. For example, it is important to know if the same group of whales is subjected to hunting in more than one location.

Besides learning more about the whales, we must look upon these magnificent creatures of the Arctic seas with awe and respect. We must take responsibility for understanding their needs and sincerely put them first. The Arctic Ocean belongs first to the Arctic whales. It is their home, their only home.

How much wiser it would be to care for an ocean wilderness before it is despoiled rather than to attempt repair after the damage has been done.

FOR FURTHER READING

Bruemmer, Fred. 1985. *The Arctic World.* Toronto: Key Porter Books.

Bruemmer, Fred. 1993. *The Narwhal: Unicorn of the Sea.* Toronto: Key Porter Books.

Jenkins, J. T. 1921. *A History of the Whale Fisheries.* Port Washington, N.Y./London: Kennikat Press.

Slipjer, E. J. 1979. *Sea Guide to Whales of the World.* 2nd English ed. Scarborough, Ont.: Nelson Canada.

Smith, T. G., D. J. St. Aubin and J. R. Geraci, eds. 1990. "Advances in Research on the Beluga Whale." *Canadian Bulletin and Aquatic Sciences* 224. Ottawa: Department of Fisheries and Oceans.

FACING PAGE: *No pristine area need be despoiled. To do so is a matter of deliberate choice. Not to do so is also a matter of choice.*

INDEX